AF622585

What Difference Does Time Make?

What Difference Does Time Make?

Papers from the Ancient and Islamic Middle East and China in Honor of the 100th anniversary of the Midwest Branch of the American Oriental Society

edited by

JoAnn Scurlock and Richard H. Beal

ARCHAEOPRESS PUBLISHING LTD
Summertown Pavilion
18-24 Middle Way
Summertown
Oxford OX2 7LG
www.archaeopress.com

ISBN 978-1-78969-317-1
ISBN 978-1-78969-318-8 (e-Pdf)

Printed in England by Oxuniprint, Oxford

Contents

Wrinkles in Time

Acknowledgements

The editors would like to thank a number of people without whom this volume would not have been possible. First and foremost, we wish to recognize the contributors who are among the easiest cats it has been our pleasure to herd and whose inspirational offerings would grace any volume. Then we owe a debt of gratitude to Stacy Davis of St. Mary's College in Notre Dame (South Bend), Indiana who was the gracious and ever efficient hostess of the 100th anniversary meeting of the Midwest Branch of the American Oriental Society, which met in conjunction with the Midwest Branch of the Society for Biblical Literature and with the blessing of the American Schools of Oriental Research. Kudos to Eric Mason, coordinator for the Midwest SBL. Also to be thanked is Laura Battini who connected us with Archaeopress. Last but by no means least, there are David Davison, the editor of Archaeopress and Ben Heaney, our compositer and cover designer, for friendly, efficient, and unbelievably speedy service for which we shall be eternally grateful.

Contributors

Laura Battini is a researcher in archaeology and history of the Ancient Near East. She is author of *L'espace domestique en Mésopotamie de la IIIe dynastie d'Ur à l'époque paléo-babylonienne*, and a guide to Syria: *La Syrie* (Éditions Peuples du Monde). She has edited 5 books on army, war, iconology of war, domestic architecture and medicine. She is chief editor of the journal قرشال Ash-sharq as well as of the series Archaeopress Ancient Near Eastern Archaeological Series, and of the scientific blog *Sociétés humaines du Proche Orient ancien* (https://ane.hypotheses.org/).

Richard H. Beal received his PhD from the Department of Near Eastern Languages and Civilizations at the Oriental Institute of the University of Chicago. He is the author of *The Organisation of the Hittite Military*. He has worked for the Chicago Hittite Dictionary Project since its inception in 1976 and is now a senior research associate. He co-edited *Creation and Chaos* with his wife of 44 years.

Maria Adele Carrai completed her Ph.D. at the University of Hong Kong, where she was Swire Scholar and a recipient of the Award for Outstanding Research Postgraduate Student for 2015-16. She was a fellow at Columbia University's Italian Academy, Princeton-Harvard China and the World Program, European University Institute of Florence, and New York University Law School. She is a sinologist and political scientist with an interest in conceptual history and history of international law. She is a recipient of a three-year Marie Curie Fellowship at the Leuven Centre for Global Governance – KU Leuven, a Fellow at Harvard University Asia Center, and an Associate Research Scholar at the Weatherhead East Asian Institute of Columbia University. She is the author of *Sovereignty in China. A Genealogy of a Concept since 1840* and co-editor of *The Belt and Road Initiative and Global Governance*.

Peter Feinman is the founder and president of the Institute of History, Archaeology, and Education, a non-profit organization which provides enrichment programs for schools, professional development program for teachers, and public programs. He received his B.A. in history from the University of Pennsylvania, a M.Ed. and MBA from New York University, and an Ed. D. from Columbia University. His interests cross disciplinary boundaries including American history, ancient civilizations, biblical history, and New York history. He is author of *Jerusalem Throne Games: Battle of Biblical Stories after the Death of David*. He is the president of the Westchester Society of the Archaeological Institute of America.

Robert Haug is an Associate Professor of Islamic World History, pre-1500 at the University of Cincinnati and runs the Middle East Studies program there. He received his PhD in the Dept. of Near Eastern Studies, University of Michigan in 2010. He is the author of *The Eastern Frontier: Limits of Empire in Late Antique and Early Medieval Central Asia*. His current research interests focus on early Islamic Iran and Central Asia.

Jeffrey P. Hudon received his Ph.D. in Near Eastern Archaeology and Old Testament Exegesis from Andrews University. He is Staff Archaeologist, Staff Research Associate and Adjunct Professor, Andrews University, as well as Adjunct Professor, Department of Religion and Philosophy, Bethel University and Adjunct Professor, School of Ministry, Grace College and Seminary. He is Administrative Director of the Tall Hisban Cultural Heritage Project and a staff member on the Tall Jalul Excavation Project and Tall Safra Excavation Project.

Suhail Laher received an MA in Religious Studies from Boston University and a PhD in Arabic from Harvard. He is an expert in Quranic Arabic language, Arabic literature and rhetoric, Islamic theology, law (*fiqh*), hadith (the chain of tradition of the Sunna), and Islamic intellectual history. He splits his energy between the Fawakih institute, which specializes in teaching Quranic Arabic, and Hartford Seminary. He served as Muslim Chaplain at MIT for over ten years, an experience that helped lead him to believe strongly in the value of both inter-religious and intra-Muslim dialogue.

Wayne T. Pitard is Professor *Emeritus* in the Department of Religion and Director *Emeritus* of the Spurlock Museum of World Cultures at the University of Illinois at Urbana-Champaign. He received his Ph.D. from Harvard University in 1982. He is author of *Ancient Damascus* (1987), and *The Ugaritic Baal Cycle, Volume II (2009),* with Mark S. Smith.

Louise M. Pryke graduated from the University of Sydney with a PhD in Ancient History, and has taught courses in ancient languages and literature at Macquarie University and the University of Sydney. In 2016, she was one of five recipients of the IAA Fund - an international award for early career scholars in the field of Assyriology. She is the author of *Scorpion* (2016), *Ishtar* (2017), and *Gilgamesh* (2019). She is a lecturer and honorary associate at the University of Sydney. Currently, she is writing a book on the cultural symbolism of turtles.

Amelia Ying Qin is Visiting Lecturer, Department of East Asian Languages and Literatures, University of Pittsburgh. She received her Ph.D. in Chinese Literature, University of Wisconsin-Madison in 2013. Her main research area is Chinese literature and culture from the Tang 唐 (618–907) through Song 宋 (960–1279) dynasties, with a special focus on the roles anecdotal narratives played in constructing memories of the past. She is currently working on her first book manuscript on anecdotal cultural memories from pre-modern times. She is also interested in the comparative study between Chinese and Western intellectual traditions.

JoAnn Scurlock received her PhD from the Department of Near Eastern Languages and Civilizations at the Oriental Institute of the University of Chicago. She is the author of *Magico-Medical Means of Treating Ghost Induced Illnesses in Ancient Mesopotamia, Diagnoses in Assyrian and Babylonian Medicine* (with Burton Andersen), and *Sourcebook for Ancient Mesopotamian Medicine* as well as co-editing *Creation and Chaos : A Reconstruction of Hermann Gunkel's Chaoskampf Hypothesis.* She is president of Midwest Branch of the American Oriental Society, past president of the Chicago Society for Biblical Research and past co-chair of the Assyria and the Bible section of the Society for Biblical Literature.

Xiaoshan Yang received his PhD from Harvard University in 1994. He is currently Associate Professor and Director of Undergraduate Studies at the University of Notre Dame. He teaches Chinese language, literature, and culture. He specializes in classical Chinese poetry and poetics. His publications include a monograph on comparative poetics as well as articles on Tang-Song poetry, prose, and political discourse. His book *Metamorphosis of the Private Sphere: Gardens and Objects in Tang-Song Poetry* (Harvard 2003) was translated into Chinese and published by Jiangsu Renmin Chubanshe in 2008.

Introduction

What Difference Does Time Make?

JoAnn Scurlock

This volume is the publication of a conference designed to commemorate the 100th anniversary of the founding of the Midwest branch of the American Oriental Society. We begin with Wayne Pitard's paper: 'The Middle West Branch of the American Oriental Society at 100' that outlines the history of the organization from its founding in January 1917 to the present day. Among other tidbits and outrages, we learn that universities even then were stingy and that, as result of the fact that meetings of the main society were invariably held on the east coast, very few orientalists in the Midwest actually bothered to join the society. This in turn resulted in a situation in which the main society's general AOS meeting rarely drew more than forty attendees.

The theme for our session was correspondingly: 'What Difference Does Time Make? We gratifyingly received papers reflecting the wide scope of the American Oriental Society in both time and space, with participation by Archaeologists (Battini, Hudon), Assyriologists (Pitard, Pryke, Scurlock), Biblicists (Feinman), Islamicists (Haug, Laher) and Sinologists (Carrai, Qin, Yang).

Our theme is divided into four sections. The first section 'Time as a Concept' grapples with the question of what time is anyway. In Amelia Ying Qin's article: 'From Beyond Time and Space: Master Zhuang's Cosmogony and Modern Physics' we are introduced to Master Zhuang, a 4th century BCE Daoist philosopher whose conceptualization of time and space bears comparison with the time-space continuum and Big Bang theory of modern physics. Throughout the paper, a Classically informed person will hear intriguing echoes of pre-Socratic philosophy, in particular Anaximander and Zeno. For an Assyriologist like myself, the most interesting part of the discussion is in the concept of a 'point' in space or time. This is an entity that 'has a location in space but takes up zero space' or is a 'zero-length moment that can be located on the time line but has no duration in itself'. Master Zuang packages Creation as there 'suddenly' being Being and Nonbeing which, in the absence of time of however short a duration, can not be distinguished from one another.

In what is known to scholarship as the Babylonian Creation Epic, the primordial gods are imagined as just such 'points' that appear suddenly by spontaneous generation, creatures neither living nor exactly dead but more or less both at the same time and located, if that is the word for it, somewhere and nowhere in infinite time and space. After the god Marduk gets through with them, however, they are points on a line moving along tracks laid out for them, being killed and brought back to life in a yearly cycle that produces and maintains the flow of time.

The second section 'Measuring Time' deals with issues of calendric ritual. Competing cultic calendars can provide real headaches for an imperial system. In his article: 'The Gifts of Mihragān: Muslim Governors and Gift Giving During Non-Muslim Holidays', Robert Haug tells a fascinating story of accommodation between the Zoroastrian population of Balkh and their Muslim Arab conquerors that involved an adaption of the Zoroastrian festival of Mihragān. This festival was a hard pill indeed to swallow, since it involved the giving of gold and silver vessels from which Muslims were not supposed to eat and silk garments which Muslims were not supposed to wear and, worst of all, getting drunk on an alcoholic beverage that was not legal even by the most liberal interpretation of Islamic law.

The difficulties experienced by Muslim rulers of Zoroastrian subjects are, to some extent, to be laid at the door of monotheistic salvation religions. When polytheism was the only kid on the block, it was possible, through creative syncretism, to combine cultic calendars for key festivals. My article: 'Feasting in the Garden of God: Ramet Rahel and the Origins of Rosh Hashanah and Yom Kippur.' traces the history of an imperial composite calendric ritual from its first attestation in the Second Millennium BCE Amorite kingdom of Mari via Assyria to Judah in the Neo-Babylonian and Persian periods.

This article pairs with Haug's article since the Mihragān festival that he describes is the final end point in Persia of the history of the ritual whose origins I have attempted to trace in my own article. From this vantage point, I believe that I can weigh in on a few issues raised in Haug's paper. The association of the Autumnal Equinox, the original date of Mihragān, with 'equity and equilibrium' and its role as a time for decision making is universal in the Ancient Near East, but never under the aegis of a solar divinity. Ninurta and Nabû, the gods who supervise pacts, contracts and covenants in ancient Mesopotamia, are the children of heads of pantheon; they are associated with the planet Mercury and, although fully involved at the Vernal Equinox, they are off the playing field in the Autumnal Equinox.

What this means is that whether you imagine Mithra as originally a son of Ahuramazda and a Ninurta and Nabû analogue or as originally a Sungod, the association with the Autumnal Equinox will have had to be purposeful rather than happenstance, if the Ancient Near Eastern evidence is any guide. Also, the association of the Autumnal Equinox with the harvest festival will have been original, with Mithra added later, perhaps as an earlier Zoroastrian accommodation to the local customs of its non-Zoroastrian subjects. I would add also, that the harvest in question is of grapes and that the special royal wine drunk at the festival will have been select vessels of new wine, as in the Ugaritic version. The placement of the festival on the 16th rather than at the beginning of the month, as is the case with Nowrūz, will have given the grape juice enough time to ferment.

The third section 'Travels through Time' deals with changes brought about by the slow but steady progress of time. Two of the papers in this section deal with changes in the life of individuals. In Louise Pryke's delightfully written article: 'Time Travel and the Homesick Hero' we hear of the way in which the homesickness that can result from extended journeys abroad 'involves more than geographical and temporal displacement, but is deeply connected to issues of humanity, morality, identity, religion and community.' Heroes whose epic journeys produce such transformations include not only the more familiar Gilgamesh, but also Adapa, Etana and especially the usually ignored Enkidu and the ultimate homesick hero, Lugalbanda.

In Xiaoshan Yang's 'The Older the Better: The Critical Conception of Lateness in Song China' we decide if there is some virtue to getting old. At my age, I am rather skeptical of this. but we can all take consolation in the canonization of Du Fu (712-770 CE) as the greatest of all Chinese poets. This process, which did not begin until four decades after his death, was accompanied with much roaring and screaming and reams of poetry about what exactly it was that was so special about Du Fu or even if he deserved canonization at all. In the end, it was decided that not only Du Fu's 'heroic abandon and plaintive cadence', but the work of later poets and even some prose writers were fine wines that only improved with advancing age and/or life experiences. I would not, however, advise mature scholars to follow the lead of Su Xun (1009-1066) and all too many others in burning their 'juvenile' works in favor of more mature pieces that aspire to be 'heroically archaic' or fashionably modern. As we hear in the conclusion, in the West as in the East, merit in a work of literature or, in the West art and music as well, is in the eye of the beholding critic.

The remaining papers in this section address the issue of gradual change within a community. In her article, 'The Evolution of Terracottas From the First to the Second Half of the II Millennium BC,' Laura Battini studies changes over time in the production of terracottas of the type ordinary folk kept in their houses. Inter alia, the later examples represent gods not as human figures but through symbols. There are also some very curious female figures with their hands over their eyes that we would love to know what that was all about.

Finally for this section, we have Suheil Laher's discussion of the effects of a perhaps inevitable 'decay of knowledge over time' among a different community of scholars: 'Through a Glass, Darkly?: Sea-Water, Bequests, and the Textualism-Tradition Tension in Islamic Law.' Not to spoil a peek into the twists and turns of argument of our scholarly ancestors from the 9th to 11th centuries of the so-called common era desperate to base their legal rulings on written sources of proven reliability, to know is that the issues addressed were whether sea-water could or could not be used for ablutions and whether you could bequeath something additional to someone who was already an heir under Islamic law.

The last section 'Wrinkles in Time' deals with dramatic political developments that bring major changes in their wake in which time seems to be speeded up and even is imagined as ending altogether. The first two papers deal with history-changing events, real or imagined and of local or more extended significance. Jeffrey P. Hudon's paper: 'Converging Lines of Evidence: Archaeological and Chronological Data Supporting a 792 BCE Date for the Iron IIA-IIB Transition in the Southern Levant' begins with a charming Ode to a Chronological Chart but then tells the sad story of the city of Lachish in Judah, a key administrative center 'surpassed in significance and size only by Jerusalem ... (that) perhaps even functioned as Judah's second capital.'

What this, of course, meant was that every passing army had to put a hole or two in its fortifications, and the final destruction by the Babylonian army marked the beginning of the exile. Before this Level II destruction, there was the damage done to Level III by Sennacherib and there were two recognizable destructions before that, one each for Level V and Level IV, the interest of this article. So who is to claim responsibility? Hazael of Damascus? Jehoash of Israel? Those who led the revolt against Amaziah? Uzziah's earthquake? This last possibility 'rests on unstable foundations' and conspirators are not in the habit of destroying the fortresses of places they intend to govern. As for Hazael, 1Kings 2.18-19 clearly indicates that he was bought off without doing any substantial damage. This leaves Johoash of Israel and both archaeological and textual evidence is mustered in favor of this interpretation. It would seem that the 'thistle of Lebanon' (1Kings 14.9) was indeed even more thoroughly trampled than we knew.

In Peter Feinman's 'The Hyksos and the Exodus: Two 400-year Stories,' we hear the tale of two historically transformative events. The expulsion of the Hyksos from Egypt is probably real, if heavily fictionalized in the versions we have of it; of the historicity of the Exodus, there has been a great deal of debate. What is curious is that both stories share a common element, the figure of 400 years for the duration of the time spent in Egypt. So what is so special about the number four? 'The number four, representing our ability to see forward, behind, to the left and to the right encompasses everything.' There is also special significance attached to forty, of which fifteen examples are cited by Feinman, and this only from the Hebrew Bible. Feinman argues, and I would agree, that this number 'represents the completion or fulfillment of a measure of time.' The appearance of this magic number in the Mesha stela raises interesting questions about cultural influences. The number of 400 or thereabouts has also had many takers, nine examples, cited by Feinman, as does 4,000 or thereabouts with seven.

All in all, the 4 series would seem to be firmly rooted in the West Semitic world. So why, we ask, does Ramses employ this 400 year figure and, just for good measure top it off with month 4 and day 4? And if the expulsion of the Hyksos was actually the inspiration for the Exodus story, did the Israelites cite the number from Ramses rather than the other way round?

The plot thickens as we examine the context of the 400 year stela more closely and compare it with the inscriptions of Ramses' father Seti I, who boasts of being at the beginning of a new era in Egyptian history and Ramses' son Merneptah, the putative author of the Quarrel Story that tells the tale of expulsion of the Hyksos from Egypt. I will not spoil the story by giving away the ending, but suffice it to say that our conceptualization of Egypt as at all times and under all conditions essentially racist on the subject of foreignors may require some modification.

The last articles are about revolutionary change. Maria Adele Carrai's article: 'Chinese sovereign revolution: temporal acceleration toward a better future?' is a fascinating study of changes in the meaning of the word revolution in both West and East to refer not to a cycle, which is of course what the word actually means, but to an endless forward motion towards a new and supposedly better future that requires an entire rejection of the past as well, one might add, in both the French Revolution and the Cultural Revolution of China that was modelled on it, an ISIS/Daeš-style destruction of the monuments of that past. In my article that finishes the volume, 'Nippur's Galileo Problem', I examine the revolution that, in ancient Mesopotamian thought, ensured the beginning and maintenance of time.

Front Cover: The photograph was taken by JoAnn Scurlock at Persepolis, Iran. The relief is located on one of the stairways to the main audience hall of the Palace of Darius I and Xerxes I below rows of tribute bearers. The scene depicts the Vernal Equinox, symbolically representing Nowrūz (Persian New Year), when the 'gifts of Nowrūz' were to be presented.
Back Cover: The photograph, also taken by JoAnn Scurlock, shows a water clock consisting of two bowls set on a tray. There is a large bowl filled with water and a small bowl that floats in it. A small hole in the bottom of the little bowl allows it to gradually fill with water, thus telling the time.

The Middle West Branch of the American Oriental Society at 100[1]

Wayne T. Pitard

The Middle West Branch of the AOS, the earliest regional section of the society, held its first meeting in January of 1917, just over a century ago. As part of our celebration of that event, it seemed appropriate that someone should examine the history of the branch, to look into the reasons for its founding, its role in the development of the parent society and its achievements over its lifetime. It was a pleasure to take on this assignment, and in the course of my research I found that the Middle West Branch played an extraordinarily important role, not generally known, in the expansion of the parent society during the 1910s and 1920s from an essentially regional society into a genuinely national organization. The branch was founded by scholars such as A. T. Olmstead, James Henry Breasted, Julian Morgenstern and others, who were dedicated to making it an active and valuable part of Oriental scholarship in the Midwest, and in spite of its ups and downs through the decades, the branch has more than fulfilled its creators' expectations.

Sources for recovering the history of the society are somewhat limited. The founders of the branch and even its members of the second and third generations have long since passed away, and if there is any surviving oral tradition about the early days of the branch, I have not heard it. In 1966 A. K. Grayson, then secretary-treasurer of the branch, took upon himself the task of gathering all the known files of the society and placing them into chronological order so that he could produce a brief history of the branch for the 50th anniversary celebration in 1967 (published in Sinor 1969—see the list of publications of the MW Branch below). For the early years Grayson was able to locate substantial records from 1922-26 and 1937-41. From 1946 to 1977 the records are relatively full (with the exception of 1958-60). Recent secretary treasurers have yet to add their files to the collection, I think because the cardboard box that contained the files of the branch, which went from secretary to secretary, finally became full in 1977! These files are the most valuable source for the general history of the branch, and can be supplemented with the published proceedings of the meetings that appeared in JAOS from the beginning until 1979. This is supplemented by a second, though smaller, trove of material among the papers of James Breasted, the father of the Oriental Institute and co-founder with A. T. Olmstead and Julian Morgenstern of the Middle West Branch, located in the Oriental Institute archives. These materials particularly shed light upon the founding of the branch and the activities of the first three meetings in 1917-1919.

I would like to focus in some detail on the first years of the branch, from the preparations for the first meeting, which was held January 27, 1917 through the meeting of 1922, when the national society met in Chicago in joint session with the MW Branch, the first time the national meeting had not been held on the east coast. I then would like to look more generally across the following decades up to today.

---

[1] A much earlier version of this paper, written by the author for the 75th anniversary of the Branch, was published as 'The Middle West Branch of the American Oriental Society at 75: A Look Back,' in the *Newsletter of the American Oriental Society* 15(1993): 5-17.

The Early Years: 1916-1921

The place of honor in the founding and nurturing of the MW Branch belongs to A. T. Olmstead, the famed ancient Near Eastern historian. The original idea of forming a Western Branch of the AOS belonged to Olmstead, who at the time was an Associate Professor of History at the University of Missouri at Columbia (Breasted to Edgerton, Dec. 15, 1916).[2] Primary among the reasons for his proposal was the fact that few scholars from the Midwest were ever able to attend the annual AOS meeting, which was always held on the east coast. In those days travel by train was laborious, and universities did not provide travel funds for attendance at conferences. Olmstead believed that the meetings were a major benefit of any scholarly society, and that there were enough orientalists in the Midwest to hold profitable annual meetings, which would encourage collegiality through the region in a way that had not occurred before. Since few of the orientalists in the middle part of the country had bothered to join the AOS, Olmstead was convinced that an accessible annual meeting under its sponsorship would help increase the membership of the society.

Olmstead broached the idea to James Breasted, who at first was reluctant to support it, since he feared that the society would be split, with no one from the Midwest bothering to attend the parent meeting and certainly no one from the east coming in the other direction! But eventually Breasted became convinced that a branch might be worthwhile, and by early 1916, he had become a very powerful proponent of the idea. I suspect that Breasted's support was a critical element in gaining acceptance for the idea at the national level. Olmstead was still a relatively young and unknown scholar, isolated from the field in Missouri, while Breasted was moving toward the peak of his career, just three years away from founding the Oriental Institute at the University of Chicago. Breasted's support obviously carried weight in the Society.

The proposal was brought to the Directors of the AOS at the 1916 annual conference in Washington D.C. and discussed in a directors' meeting on April 25. The directors unanimously voted to appoint a committee consisting of Breasted (chair), Olmstead, Julian Morgenstern and Albert Clay (the secretary-treasurer of the AOS) to look into the advisability of forming the branch. The committee proposed that a meeting be called of all interested orientalists during the winter of 1917 to see whether there was enough interest to make it work.

It is clear from the correspondence that passed between Breasted, Olmstead and Morgenstern (Clay played a more peripheral role as a representative of the east), that they were dedicated toward making the plan a success. Breasted himself undertook to organize the program for the meeting, which would be held at the University of Chicago. He wrote numerous personal letters to scholars throughout the Midwest, encouraging them to present papers at the meeting and to join the AOS. The letters had a good effect. A. H. Lybyer of the University of Illinois in Urbana, a specialist in Turkish history, responded on November 6, 1916 to an invitation from Breasted saying, 'I am pleased to hear of the proposed Western section of the American Oriental Society. The fact is that I had become discouraged, and at the time of paying my dues for the current year, sent in my resignation. I have received no acceptance of this, and am willing to continue now that there is a prospect of the society's possessing more meaning for me than the occasional arrival of the Review.' Lybyer was to become a long-term supporter of the branch.

2 Unless otherwise noted, the unpublished materials cited in this section all belong to the Oriental Institute Archives.

Olmstead and Morgenstern did what they could as well. Initial fears that no one would show up for the meeting quickly dissipated. In a letter dated Dec. 10, 1916, Morgenstern informed Breasted that he would bring an invitation from HUC to hold the second meeting of the branch in Cincinnati, in spite of the fact that the branch was yet to be formed!

In his final pre-conference letter to Olmstead and Morgenstern, Breasted informed his fellow committee members (Jan. 24, 1917): 'You will be glad to know that we have about forty acceptances, and unless many disappoint us, we shall have as large a meeting as is customarily held in the East.' And indeed the meeting was a great success. Forty-five members (including 26 new ones) attended the meeting, and additional local attendees brought the figure to over 75. That there was a clear interest in a Middle West Branch could not be doubted. The eastern leadership was impressed.

At the meeting the first officers were elected: Breasted as President, Dr. Bertholdt Laufer of the Field Museum as Vice-President, and Olmstead as secretary-treasurer. Olmstead was a bit reluctant to take on his new position, as he felt he knew so few people in the field (Olmstead to Breasted, Feb. 14, 1917). But he performed the job with extraordinary dedication for the next four years.

The second meeting was held at Hebrew Union College, and it went quite smoothly after difficulties about meeting on a Saturday were ironed out in a fascinating exchange of letters between Morgenstern, Breasted and Olmstead. As one might imagine, the faculty of HUC found it inappropriate to hold the meeting on the Sabbath, but Breasted and Olmstead feared that no one would come during the week. Morgenstern solved the problem by proposing that the meeting be held on Friday, February 22, Washington's birthday, which was a holiday on all campuses at that time. The attendance was considerably smaller, only 21 members, but Olmstead noted in his account of the proceedings that more out-of-town members had come to this meeting than to the first one. He also noted that Professor Breasted's Presidential Address had drawn a crowd of some 150 persons.

The third meeting was a disappointment for Olmstead, who had moved to the University of Illinois at Urbana in 1917. Hopes for a large and successful conference there faded because of the horrific influenza epidemic of the winter of 1918-19. Only 16 members made it. But the following year, 32 members were at the meeting at Northwestern University and Garrett Seminary. In 1921 the meeting was again small, with 17 gathering at University of Wisconsin in Madison.

The Joint Meeting of 1922 and its Aftermath

Even before the 1921 meeting was held, the leaders of the branch had decided to try what, for the time, was a rather daring proposal--to invite the parent society to hold their 1922 meeting in Chicago in joint session with the Middle West Branch. This was unprecedented, since the AOS had always met on the east coast, between Boston and Washington. Breasted and Olmstead again seem to have been the major players in this project. The first reference to the plan that I have found occurs in a letter from Olmstead to Breasted, dated Jan 24, 1921, where he mentions having already discussed it with Breasted. On January 26, Breasted wrote Olmstead that he was hoping to get both the Field Museum and the University of Chicago to issue a joint invitation to the parent society, and by February 14, he had also gotten the Art Institute as an additional host. Olmstead responded happily, 'Won't we show the brethren from the east what's what,--if anybody comes. Perhaps a select few may venture out to have a chance to visit the Indians.'

The invitation was sent to Charles Ogden, the Corresponding Secretary of the the AOS on March 17, 1921. Although I have no records of the discussion among the eastern members, it is clear that when the invitation was brought up at the annual meeting, there was great controversy about it. In 1927, Ogden, still corresponding secretary, wrote (March 22) to the Middle West secretary, Ovid R. Sellers, 'The reference in your circular letter to 'the reluctance of some of the Eastern members to hold a meeting in the Middle West' is probably a reminiscence of the conditions prior to the first joint meeting in 1922, when some prominent members were opposed to such a step.' In spite of the controversy, however, the executive committee voted to accept the invitation.

The correspondence relating to this meeting, both from the MWB files and from Breasted's papers, is extremely animated and interesting. The new secretary, George Allen of the Oriental Institute, Breasted and Olmstead made every effort to get Midwestern scholars to offer papers, stressing the importance of making a good show. To one member, Allen writes, 'If we mean to make the right impression on skeptical Easterners we must be present and also take part (underlining his).'

The paper proposals were slow to come, and the leaders were very anxious at times. On March 22, a week before the deadline for printing the final circular for the meeting, only three midwesterners had submitted papers. Olmstead wrote to Breasted on March 22, 'Doctor Ogden is much worried about the program. I have been so taken up with my family [his daughters had had bouts with scarlet fever--Olmstead to Allen, Feb. 28] that I have neglected to keep in touch with the meeting as I should otherwise have done. Probably I did not indicate to Doctor Allen how necessary it has been to camp on everybody's trail, until they promised to send in a paper. The result is that with the time for appearing on the program supposed to be reached, just three persons, Professors Smith, Sprengling, and Morgenstern, from our branch have promised papers, while ten have done so from the east. It puts us in a pretty bad light, and Ogden makes it pretty plain. The only thing I see to do is to write everybody who is ordinarily on the program a personal letter. I am accustomed to put in two or three names in the proof and the delays in the past were as exasperating as could be, but this is the limit.. . . For my part, if we do not get some men from the Middle West on the program, I shall never have the face to attend another meeting of the parent organization.' But within days, fourteen more members had submitted papers, and the branch was vindicated at the Meeting.

The 1922 joint meeting in Chicago was a success. Forty-eight members attended, roughly comparable to attendance at meetings of the previous decade. Olmstead wrote to Allen (May 23, 1922): 'In spite of all the bad luck croakers, the Chicago meeting was a success, but didn't we have to hustle! Of course, the east did pretty badly, but I am pleased at the way our people turned out and read papers. And if I do say it myself, our papers mostly had point and pep. The easterners who did come heard and saw some things to take back. And as a neutral, it did me good to see how the University of Chicago rubbed it in that it was the leading oriental institution in the country. It was worth the fare to Chicago to see that fact gradually sinking into a certain Haupt.'

The events of 1917-1922 are, of course, significant to those who are interested in the MW Branch, but they are also of exceptional significance for the history of the parent organization as well. To see this one must examine the membership lists of the AOS. In 1917, the American Oriental Society was for all intents and purposes a regional organization of the Northeast. The 1915 list of members contained 304 regular members of the society, of which 39 (or 12.5%) lived off the east coast, but not west of the Rockies. Only four members lived in the far west, four were in Canada, and 25 (or 8%) were scattered in various foreign countries. On the other hand, 232 (or 76%) lived on the east coast. One of the key purposes of the formation of the branch was to increase the membership of the AOS in the Midwest (described by Olmstead as 'everything from the Rockies to the Alleghenies.') And the efforts of the Middle West leadership paid off handsomely. At the 1921 meeting, Olmstead was able to report that there were now 116 members in the branch, and by 1922, Ogden assigned 153, a bit over 25% of the society to the MW Branch, while the east coast membership now made up only about 60%. It becomes clear then that the Middle West Branch played a primary role in transforming the AOS into a truly national organization for the first time.

The joint meeting in Chicago in 1922 was also an epoch-making event for the society as a whole and an important signal of the nationalization of the society, as it decisively broke the meeting away from its perpetual northeastern mooring. The second Midwestern meeting of the national AOS occurred just five years later, at Cincinnati, and Ogden wrote to the Midwest secretary in March of 1927, 'As I remember, no one at Philadelphia last year demurred at accepting President Morgenstern's invitation; and, though some of the Eastern members have written to me regretting that the meeting this year is too far away for them to attend, I think that no one has expressed disapproval of the principle of a joint meeting. I hope therefore that we can consider the precedent safely established.' After that, joint meetings were held in the Midwest in 1930 (Toronto), 1932 (Chicago), 1935 (Ann Arbor), 1937 (Cleveland), and 1941 (Chicago).

One can also see this nationalization apparent in the increasing participation of midwestern scholars in the offices of the AOS. It is quite clear that an underlying reason for the founding of the branch, not specifically stated in the correspondence, was that the eastern society had largely ignored its members in the west. There clearly is some regional tension evident in the correspondence. As the size of the midwest branch grew, it became increasingly important to the leadership that the interior of the country be represented more substantially in the national organization. Interestingly, however, there was concern early on that pressing this issue might create bad feelings among the eastern establishment. In 1918 Olmstead was elected to the Board of Directors, but Olmstead to reluctant to accept. He wrote to Breasted (April 11): 'I have written much to the men of the east in regard to fuller representation in the parent organization but I have also constantly insisted that I had no personal desires in the matter. I know they meant it for the best, but it does look as if I were playing for my own hand. I think the danger of misunderstanding would not be confined to the parent organization. I have already heard rumors that we of the organization committee were fixing all the wires and taking all the offices, and this would seem to prove it to the hilt. I think under all circumstances, I ought to be permitted to resign and to have a successor chosen. So far, we have not played politics, but I am almost persuaded we ought to take more care to see that the offices are more widely distributed.' Breasted responded (May 1, 1918): 'I am glad to have your hint regarding impressions anywhere that we Westerners are endeavoring to stampede the Oriental Society. Certainly your attitude regarding the directorship ought to disabuse their minds if any such impression has arisen. Of course Eastern leadership of the Society has been such an inevitable matter of course for so many years that there would be no harm done if Western members were given fuller representation than has heretofore occurred, but I should hate to have any one think that we were playing politics to bring this about, and I should much rather see things go on in the old way than to produce any such impression.' In March of 1922, Olmstead wrote to Allen (March 29, 1922): 'We of the Middle West have a right to more representation. It is a simple scandal that we do not have our secretary ex officio a member of the board of directors. We cannot do it this year [the national meeting was being held in Chicago] or it will look as if we had put one over on them, but I am going to serve notice that next year you are to be put on, if it means WAR! I made a count the other day. We have a trifle over twenty five percent, the easterns, including Canada and the south, a little less than sixty percent. We should, on numbers, have five to twelve. I do not think we are inferior in productive scholarship.. . . I shall write to Jackson (chairman of the nominations committee) about this, not, of course in the same language.' This movement thus led to the diversification of the society in new and advantageous ways.

The 1930s to the 1970s

The Middle West Branch steadily grew through the twenties and flourished in the 1930s, when meetings that were not joint with the parent society attracted increasing numbers from the lower to mid thirties early in the decade, to 81 at the 22nd meeting in Chicago in 1938.

The advent of World War II naturally brought about a diminution of the meeting, but one momentous development occurred during those years. In 1942, twenty-five years after its founding, the Middle West AOS met jointly for the first time with the Midwest Section of the Society of Biblical Literature. That, of course, proved to be a very happy union and one that has continued almost without interruption since.

After the war attendance never again reached the levels of the late 1930s. Certainly by this time, transportation had made it feasible for anyone to attend the national conference, and since the branch meeting was normally held within a few weeks of the national, attendance at both was often difficult. Through the 1950s and 1960s, AOS attendance at the meetings tended to be in the upper twenties to lower thirties when held outside Chicago, upper thirties to lower forties when in Chicago.

In 1966 a committee was formed to look into the issue of whether the Middle West Branch had outlived its purpose and should perhaps be dissolved. The committee decided that although the original reason for founding the society was no longer an issue, the branch still had some important functions, such as allowing scholars to meet and talk in a more informal setting than at the national meeting and providing younger scholars and graduate students a forum in which to present papers. In 1967 the branch celebrated its fiftieth anniversary with great enthusiasm and published a commemorative volume of essays in honor of the semi-centennial, the first book-length publication directly resulting from a Branch meeting.

The Middle Western Branch had always been dominated by the Ancient Near Eastern component of the society, both because of the influence of the Oriental Institute of Chicago and because that is the region of Asia of most interest to the members of the Midwest SBL. In the 1960s there were still sessions on South and East Asia, but by the mid-seventies these had largely vanished.

The Middle West Branch and Major Symposia: 1979 to the Present

An important new phase of the Middle West Branch began in the late 1970s. Although the number of active members was relatively small, those involved were an ambitious group that decided to use the Middle West AOS meeting to hold international symposia on subjects of common interest to the largely Ancient Near Eastern AOS members and those of the Midwest SBL. The first large symposium with speakers from Europe and across the US was held in 1979, celebrating the 50th anniversary of the first expedition to Ugarit, Syria. It was a substantial success and led to the publication of the Branch's second edited volume, *Ugarit in Retrospect: Fifty Years of Ugarit and Ugaritic* in 1981 by Eisenbrauns. This encouraged the group to organize a second symposium, this one on Mari at 50 in 1983, which resulted in another volume in 1992; and then the largest of the symposia, celebrating the centennial of the Amarna Letters, in 1987 at the Oriental Institute. The latter brought 329 scholars to the meeting, the largest gathering in the joint Midwest AOS/SBL's history. Fifty-nine papers on Amarna were read at the conference over its packed three days.

These were succeeded by additional symposia on Emar (1991); a special set of studies in honor of Michael C. Astour (1995); daily life in the Ancient Near East (2001); on Ugarit at 75 (2003); and on Hermann Gunkel's Chaos Kampf Hypothesis in 2011. All of these symposia resulted in published volumes. Thus between the years of 1966 and 2013, eight volumes were published (a ninth, the Amarna Volume, was scuttled by the difficulties of getting all the papers to the editor in a reasonable time period) from the Middle West Branch's symposia held at the annual joint meeting. The significance of the research presented and published from the Middle West Branch is unprecedented--I do not believe there is any regional branch of either AOS or SBL that has even approached this record of achievement.

Over the past century, the Branch has had a large number of illustrious scholars who served as President. The list of presidents found below is a testimony to the important contribution of scholars who have been located in the Midwest. But it is also important to say a word about the office and officers who in many ways have been the true backbone of the society, who were the real workhorses and whose labors have been a major element in the success of the meetings of the branch-- the secretary-treasurers. There have been fewer than 25 over the past 100 years. Before the first meeting in January of 1917, Breasted wrote a letter to Olmstead and Morgenstern, in which he described the two offices he thought the branch needed: 'It might be sufficiently explained to name the offices,--presumably a president and a secretary-treasurer. The latter official would have all the work to do, or most of it.' And this has been the job description for the secretary-treasurer ever since. The copies of correspondence that make the file folders in the archives bulge is testimony to the dedication of the secretaries through the years. In more recent years that could be supplemented by the massive numbers of emails the secretary sends. It is a busy, often tedious job, and the early correspondence shows that it was from the beginning. For example, Olmstead, the first secretary, for whom I have developed a great deal of affection through reading his delightful letters, once complained to Breasted (Jan. 24, 1921), 'I have been writing letters all day long, mostly to people who almost certainly will not come, but like to be invited just the same, and it gets rather tiring.' When the second secretary, George Allen, was trying to get J. M. P. Smith to agree to becoming president of the branch in 1922, Olmstead, now president, advised Allen on this way to approach Smith: 'Tell Professor Smith, if he balks at the presidency of the western branch, that it is nothing, that he will get all the honor and that you will do all the work, leaving him powerless. Tell him I said so, and I ought to know, for I was secretary four years, and a powerless president one.' By his fourth year as secretary, Olmstead was convinced that the program ought to be done by a committee, and argued for that idea. But nothing ever came of it. The secretary always did most of the work. Thankfully, for the past seventy-five years the burden has been shared with the secretary-treasurers of the Midwest SBL, and this has made matters easier, as each one has taken an alternate year to be in charge of the plenary programs. I'm sure they could tell similar stories.

The Middle West Branch of the American Oriental Society still faces challenges. Its attendance is smaller than it ought to be (a perennial problem), and it fails to draw many scholars from fields outside the Near East. But over this century it has provided an extraordinary place to bring colleagues together to discuss important issues of history, language and religion and provide a less formal atmosphere in which to get to know one another. I believe that the achievements of this Branch would have gratified its founders and made them proud.

Publications of the Middle West Branch of the American Oriental Society

1. *American Oriental Society, Middle West Branch, semi-centennial volume; a collection of original essays.* Ed. Denis Sinor, Indiana University Press, 1969.
2. *Ugarit in Retrospect: Fifty Years of Ugarit and Ugaritic,* ed. Gordon Young, Eisenbrauns, 1981.
3. *Mari in Retrospect: Fifty Years of Mari and Mari Studies*, ed. Gordon Young, Eisenbrauns, 1992.
4. *Emar: History, Religion and Culture of a Syrian Town in the Late Bronze Age.* Ed. Mark Chavalas, CDL Press, 1996.
5. *Crossing Boundaries and Linking Horizons: Studies in Honor of Michael C. Astour on His 80th Birthday,* Ed. Gordon Young, Mark W. Chavalas and Richard E. Averbeck. CDL Press, 1997.
6. *Life and Culture in the Ancient Near East.* Ed Richard E. Averbeck, Mark W. Chavalas and David B. Weisberg. CDL Press, 2003.
7. *Ugarit at Seventy-Five.* Ed. K. Lawson Younger, Eisenbrauns, 2007.
8. *Creation and Chaos A Reconsideration of Hermann Gunkel's Chaos Kampf Hypothesis.* Ed JoAnn Scurlock and Richard Beal. Eisenbrauns, 2013.

Time as a Concept

From Beyond Time and Space: Master Zhuang's Cosmogony and Modern Physics

Amelia Qin

Introduction

For every being or event that has a start and an end, its existence is defined by and subjected to the constraints of time. What, then, is time? Is it independent of the physical world? Or did it begin with the world and will also end with it? Judging by its functions, time is a dynamic concept used to describe sequences, durations, intervals, and changes. There is time because there are always changes and movements in the world: living things growing old, seasons alternating, stars parting with each other at enormous speed, the universe expanding, human beings constantly thinking... Had there been absolutely no movements, no changes in any part of the universe, absolutely no difference from one moment to the next, the concept of time would become meaningless—what is a moment when there is no way to even tell it apart from the next? Time, then, is a concept that has an ultimate dependence upon differentiation.

It, in turn, naturally becomes the defining boundaries of any limited existence. By marking the changes of coming into being and dying out, it defines the duration of existence in a way very similar to that of the spatial boundaries marking the points where a body starts and ends to take up space. As is almost always the case in ancient Chinese thoughts, the concept of time in the *Zhuang Zi* 莊子, a fundamental text of Daoist thought named for its attributed author Zhuangzi 莊子, Master Zhuang (ca. 4th century B.C.), is also closely related to the concept of space. Focusing on a passage of regressive reasoning in 'Discussion on Making All Things Equal' 齊物論, the second of the Inner Chapters of the *Zhuang Zi* text, this paper discerns a cosmogony of intertwined time and space in the work of Master Zhuang. What happens at the very beginning of the cosmos? What is before the beginning? Is there an end to this regressive inquiry? By identifying explicitly the inseparable entity of time and space as a space–time complex in the *Zhuang Zi* passage, this paper is able to discuss the *Zhuang Zi* cosmogony in relation to the Big Bang Theory and understand the nature of the Dao 道, the fundamental idea of Daoism and the origin of the *Zhuang Zi* cosmos, from the perspective of time and space, or rather from that beyond time and space. It is a perspective no longer based on differentiation, and the concepts of time and space no longer apply. From this perspective, *Zhuang Zi*'s worldview that all things should be deemed equal, in all aspects of existence and non-existence, clearly is originated from and rooted deeply in its cosmogony. Viewing all things as equal is a natural result of becoming one with the Dao and returning to *the* origin of all things and the cosmos, which is beyond the constraints of time and space. For human beings with limited existence, in terms of both time and space, to return to the Dao and regard all things as equal is the Daoist way to escape the space–time limitations of their lives and enter the boundary-less realm of eternity and infinity. By viewing time and space as one unity, treating the lifespan of a man and the body of the man as one idea, passages in 'Zhi Travels North' 知北遊, the last of the Outer Chapters, and 'The Great and Venerable Teacher' 大宗師, the sixth of the Inner Chapters in the *Zhuang Zi,* demonstrate explicitly how time is understood in terms of space and that the concept of time is inseparable from that of space in ancient Chinese thought. Thus, transcending the boundaries of space allows one to transcend the boundaries of time as well; 'leaving one's form' 離形 allows one to escape the mundane reality confined by both space and time, enter the realm of infinity and eternity, and become one with the Dao.

Understanding the Dao

The origin of the *Zhuang Zi* cosmos, the Dao, is described as follows in the 'The Great and Venerable Teacher' chapter:

> 夫道，有情有信，無爲無形。可傳而不可受，可得而不可見。自本自根。未有天地，自古以固存。神鬼神帝，生天生地。在太極之先而不爲高，在六極之下而不爲深，先天地生而不爲久，長於上古而不爲老。
>
> Now the Dao has its reality and its manifestation, but is without action or form. It can be transmitted but cannot be owned; it can be understood but cannot be seen. It is its own source, its own root. Since even before heaven and earth were formed, it has been the absolute existence from antiquity. It gave spirituality to the spirits and gods; it gave birth to heaven and to earth. It exists beyond the Supreme Ultimate, and yet does not appear high; it exists beneath the Six Extremities, and yet does not appear deep. It was born before heaven and earth, and yet does not appear ancient; it is earlier than the highest antiquity, and yet does not appear old.[1]

This is a description based on a reference frame in terms of time and space. The spatial references of height, depth, and distances in four directions are represented by the Supreme Ultimate and the Six Extremities; the temporal references of duration and ancientness are the life of the cosmos and the age of highest antiquity. The Dao, however, is higher, deeper, earlier, and longer-lasting; in other words, it goes completely beyond the ranges of such a spatial and temporal reference frame. Is it because there are physical limits to the ranges of the reference frame that they fail to define the existence of the Dao? Why, then, at the same time, does the Dao not appear high, deep, old, or long-lasting at all? Or is the Dao simply a different reality that cannot be contained in such a reference frame of concrete concepts? Guo Xiang 郭象 (ca. 252–312), the most influential commentator of the *Zhuang Zi*, explains this description as follows:

> 道無所不在。故在高爲無高，在久爲無久，在深爲無深，在老爲無老，無所不在，而所 在皆無也。
>
> There is no place where the Dao is not. Therefore being high it is of no height; being long-lasting it is of no duration; being deep it is of no depth; being old it is of no age; there is no place where it is not, while everywhere it is it is not there.

Clearly, the existence of the Dao—its height, depth, duration, and ancientness—do not comply with the common understandings of time and space. The common measurements of time and space cannot capture the existence of the Dao, as it has no height, no duration, no depth, no age to be measured.

1 Translation with reference to Burton Watson's *Chuang Tzu: Basic Writings,* trans. Burton Watson (New York, NY: Columbia University Press, 1964), 77; and Wang Xianqian 王先謙, *Zhuang Zi jijie* 莊子集解 (Taipei: Shijie shuju, 1900), 40.

Cosmogony and Modern Physics

It is possible to further understand the Dao in the *Zhuang Zi* in relation to the cosmos that is defined in modern physics by space and time. First of all, for any kind of existence, time is a fundamental concept that describes the beginning and the end, as well as the changes and durations of various states of existence. There is, then, a sense of paradox when the definition of time in modern physics in fact relies on a particular change of state in one particular existence of the physical world. That is, the current definition of the second, the basic unit of time, is based on nothing but the most simple mathematics, counting to be exact, and the repetition of an event with a near-absolute, fixed duration.[2] Historically, Newtonian physics regarded time as the fourth dimension of the universe, the other three being the dimensions of the Euclidean space. Time was treated as fundamental, universal, and constant, independent of the state and movement of objects as well as that of the observer. It was without beginning or end. Thus, the Newtonian universe was a universe of absolute time and space, with an absolute, unchanging spatial–temporal reference frame. Modern physics, with the Special Theory of Relativity and General Theory of Relativity, replaced the absolute reference frame with the concept of the space–time continuum and brought time and space into an inseparable entanglement where the nature and measurement of one is interwoven with that of the other. Time is now no longer absolute and independent, but now relative, and so is space. The issue of the origin of the universe then comes up. Discussing the role of time in the Big Bang Theory, Stephen Hawking states that either time originally began with the Big Bang, or the information of events before the Big Bang belonged to a completely different timeframe and can no longer reach and have effect upon the present timeframe. Thus, time may as well be deemed as actually beginning with the Big Bang.[3]

In the chapter 'Discussion on Making All Things Equal' of the *Zhuang Zi*, there is a passage on the formation of the cosmos based on a line of regressive reasoning. Interestingly, the cosmogony revealed in the passage is, in a sense, quite similar to the origin of the universe in the Big Bang theory. The passage is translated below:

> 今且有言於此，不知其與是類乎，其與是不類乎？類與不類，相與爲類，則與彼無以異矣。雖然，請嘗言之。
>
> 有始也者，有未始有始也者，有未始夫未始有始也者。有有也者，有無也者，有未始有無也者，有未始夫未始有無也者。
>
> 俄而有無矣，而未知有無之果孰有孰無也。今我則已有謂矣，而未知吾所謂之果有謂乎？其果無謂乎？
>
> Now I will make a statement here for the time being. I don't know whether it belongs to the same category with other statements, or it does not belong to the same category with other

[2] National Institute of Standards and Technology (NIST), 'Definitions of the International System of Units: Base Units' (NIST, 2000), accessed Sept. 18, 2016, http://physics.nist.gov/cuu/Units/current.html. 'The second is the duration of 9,192,631,770 periods of the radiation corresponding to the transition between the two hyperfine levels of the ground state of the cesium 133 atom.' This definition for the second was adopted in 1967.

[3] Stephen Hawking, 'The Beginning of Time,' lecture at the University of Cambridge (1996), accessed Sept. 18, 2016, http://www.hawking.org.uk/the-beginning-of-time.html. 'Since events before the Big Bang have no observational consequences, one may as well cut them out of the theory, and say that time began at the Big Bang. Events before the Big Bang, are simply not defined, because there's no way one could measure what happened at them. This kind of beginning to the universe, and of time itself, is very different to the beginnings that had been considered earlier.'

statements. Belonging or not, combined together they become one category, then it has nothing different from others. Though it is like this, let me still try to state it:

(1) There is the Beginning. (2) There is the not yet beginning to have the Beginning. (3) There is the not yet beginning to have that not yet beginning to have the Beginning. (4) There is Being. (5) There is Nonbeing. (6) There is the not yet beginning to be Nonbeing. (7) There is the not yet beginning to be that not yet beginning to be Nonbeing.

Suddenly there is Being and Nonbeing. But I don't know, between Being and Nonbeing, which is actually Being and which is Nonbeing. Now I have just said something. But I don't know whether what I have said has actually said something, or it actually has said nothing.[4]

The central part of the passage is a set of seven propositions, all of them existential claims beginning with 'there is.' The meaning of this carefully constructed cluster of claims, however, is veiled by its succinct, almost cryptic language and has been interpreted from various perspectives by commentators throughout history.

Guo Xiang's interpretation of the seven propositions, as pointed out by Charles Le Blanc, is epistemological. For him they are 'the way man understands time, Being and Non-being and frees himself from all limitations and dualities by obliterating conceptual consciousness.'[5] In the first three propositions, the 'obliteration' 忘 of the finitude of time and of life and death, to Guo Xiang, is done through the 'equalization of the self' 自齊 with the infinite. The fourth proposition, 'there is Being,' represents the 'knowledge-consciousness' 知 of Being; the fifth to the seventh mark the progression from the knowledge-consciousness of Nonbeing to the obliteration of knowledge-consciousness itself 忘知.[6] Grouping the seven propositions into two categories, with propositions one to three dealing with the self and four to seven dealing with knowledge-consciousness, Guo Xiang's interpretation provides a way of transcending the epistemological boundaries of the self and knowledge-consciousness.

The seven propositions can also be interpreted as the description of a cosmogonic process. For example, the *Huainan Zi* 淮南子, a philosophical text on a wide range of topics compiled around 139 B.C., quotes these seven propositions from the *Zhuang Zi* for this purpose at the beginning of its second chapter, 'On the Beginning of Reality' 俶真訓. Le Blanc points out that the *Huainan Zi* commentary of the seven propositions 'has nothing to do with epistemology or logic, neither with formal or a priori concepts'; it is a kind of 'descriptive cosmogony.'[7] It involves a cosmogonic scheme of seven stages that correspond in a one-on-one relationship to the seven propositions in the *Zhuang Zi*. Le Blanc identifies seven stages in the *Huainan Zi* explication of the *Zhuang Zi* propositions as 'empty nonbeing' 虛無, 'empty and opaque space' 虛霩, 'empty and luminous space,' 'the space-time complex' 宇宙, 'blending of heavenly *qi* 氣 and earthly *qi*,' 'the life impulse' 欲生興, and 'the birth and flourishing of the ten thousand things' 萬物生興.[8] Le Blanc reordered the seven *Zhuang Zi* propositions to better match the

4 Revised from Burton Watson's translation, *Chuang Tzu: Basic Writings*, 38. Watson translates the *shi* 始 in proposition one as 'a beginning.' This paper, however, posits it is a more specific reference to *the* beginning of the cosmos in the *Zhuang Zi* cosmogony. It is translated here as 'the Beginning.' The current translation also considers commentaries in Wang Xianqian, *Zhuang Zi jijie*, 12-3.
5 Charles Le Blanc, 'From Ontology to Cosmogony: Notes on Chuang Tzu and Huai-nan Tzu,' in *Chinese Ideas About Nature and Society* (eds. Charles Le Blanc and Susan Blader; Hong Kong: Hong Kong University Press, 1987), 119.
6 Ibid., 119-20.
7 Ibid., 128.
8 Translation by Le Blanc. Ibid., 127-8.

cosmogonic scheme in the *Huainan Zi*, but he had trouble achieving a sequential cosmogonic order, because the seven propositions and cosmogonic stages do not represent 'the successive stages of the formation of the universe.'[9] It is, however, worthwhile to note that Le Blanc's cosmogonic order shows the fourth proposition, 'there was Being,' as the end of the process—the completion of the formation of the cosmos.

Inspired by both Guo Xiang and *Huainan Zi*'s understandings of the structure of the *Zhuang Zi* passage, this paper recognizes a perfectly symmetric structure in the seven propositions, and taking both time and space into consideration, identifies a sequential cosmogonic order in the passage. With the fourth proposition, 'there is Being,' as the centering point, propositions one to three and propositions five to seven form two corresponding groups parallel to each other in both grammatical construction and the progression of reasoning. The fourth proposition, 'there is Being,' claims the existence of the cosmos; propositions one, two, and three are progressive descriptions of the origin of the cosmos from the perspective of time; and propositions five, six, and seven are regressive descriptions of the origin of the cosmos from the perspective of space, as is shown in Figure 1 below. Propositions three and seven, combined together, depict the first stage in the *Zhuang Zi* cosmogony, which, as will be explained later, is called the 'boundary-less realm'; propositions two and six depict the second stage, which is before the cosmos; propositions one and five depict the third stage, which is the very beginning of the cosmos; and the last stage, the cosmos in existence, is depicted by proposition four.

In the fourth proposition, the existence of the cosmos is expressed as 'there is Being.' Two basic attributes of Being 有 are, first, a lifespan, with a starting point when the entity comes into being, and second, a presence established through taking up space, that is, literally, there 'is' 有 something. Thus the existence of Being is an inseparable combination of intertwined time and space, a space-time complex similar to the idea of the space-time continuum in modern physics. In the fourth proposition of the Zhuang Zi passage, and the fourth stage of the cosmogony embedded in this passage, the Being is the cosmos that has been lasting for a certain amount of time and is taking up a certain amount of space. With the Six Extremities as its spatial range and the highest antiquity as its temporal range, the cosmos is the reference

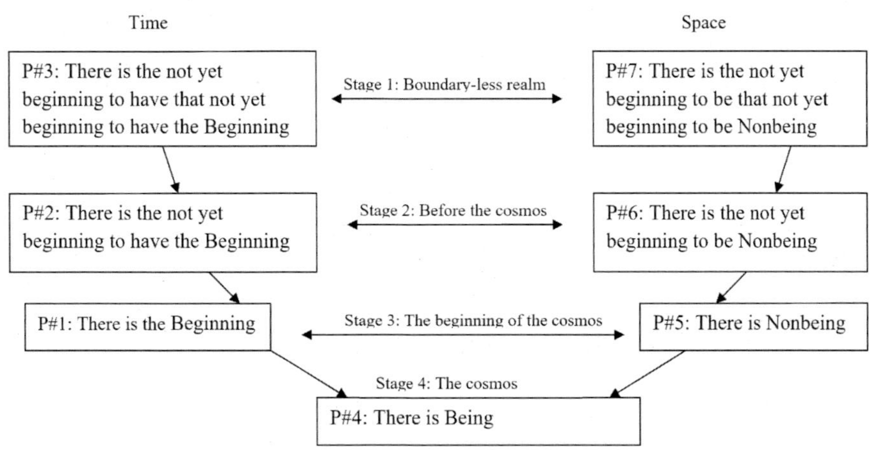

Figure 1. Symmetrically structured propositions depicting the cosmogonic process in Zhuang Zi

[9] Ibid., 126.

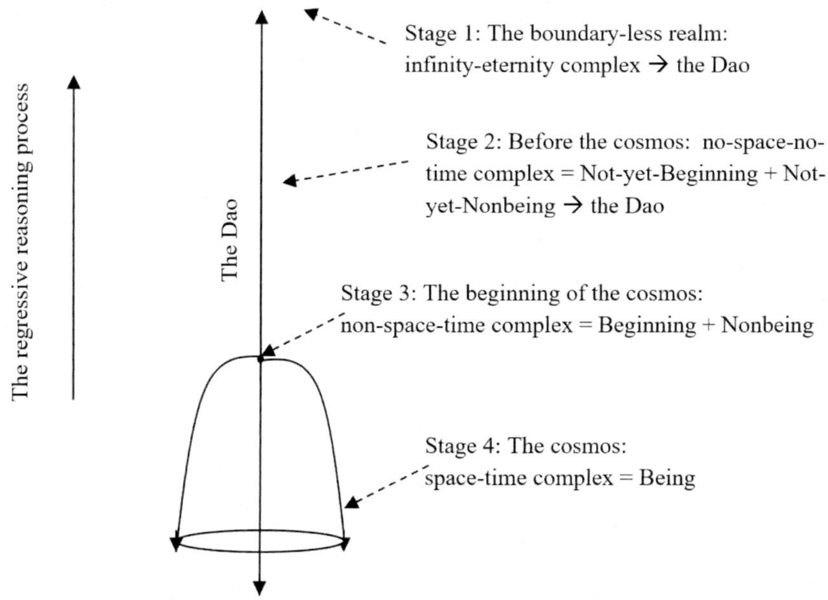

Figure 2. Zhuang Zi's *cosmogony depicted in terms of space-time complex*

frame used, albeit one insufficient for the task, to describe the Dao in 'The Great and Venerable Teacher.' The Zhuang Zi cosmogony and the four stages in terms of the combination of time and space are shown in Figure 2.

The second to the last stage (Stage Three) in the *Zhuang Zi* cosmogony, the very beginning of the cosmos, is a unity of the Beginning and the Nonbeing, whose existence are claimed in propositions one and five, respectively. A beginning, by definition, is the point in time when something starts. A point, by its mathematical definition in geometry, is an entity that has a location in space but takes up zero space—that is, without any volume, area, or length at all. A 'point' in time is a zero-length moment that can be located on the time line but has no duration in itself—that is, without either start or end. In the first proposition, the Beginning is such a 'point' when the cosmos comes into being and when the cosmos's time starts. The Beginning has to be a 'point,' otherwise it will have a duration and thus have its own beginning, which should then be considered the real Beginning.

Similarly, the Nonbeing in proposition five is the start of the cosmos's space. It should be a literal, spatial point in the sense of geometry. The Nonbeing is a stage of existence that starts to come into being but does not take up space yet. Otherwise, had the Nonbeing taken up any space at all, it would then be something existing in space; it would no longer be Nonbeing, but Being itself. Since the Beginning has no duration, and the Nonbeing contains no space, the unity formed by Nonbeing and Beginning is actually a non-space-time complex, a 'point' stage in the cosmogonic process that precedes the space-time complex of Being. However, because Nonbeing is the start of Being, Beginning is the start of time; the non-space-time complex is an infinitely small 'point' at the start of the space-time complex, empty of time and space, and perhaps only an intention, or a hint, of taking up time and space. Thus, at Stage Three of the *Zhuang Zi* cosmogony, the very Beginning of the cosmos, the start of time is ready to become time itself and Nonbeing is ready to become Being. It is a stage, just as the *Zhuang Zi* passage describes, that 'suddenly there is Being and Nonbeing. But I don't know, between Being and Nonbeing, which is actually Being and which is Nonbeing' 俄而有無矣，而未知有無之果孰有孰無也. For human perception and understanding to tell the two stages apart, there must be differences and changes that must be manifested over time, however small its duration is. It is hard to differentiate the stages of Nonbeing and Being exactly because of the 'suddenness' of the very origin of the cosmos, because the starting point of 'Beginning' does not take up any time duration, and, without time, changes and differences are not manifested to human understanding and are lost in obliteration in the non-space-time complex.

Although the Nonbeing at the Beginning, the non-space-time complex, is not a solid, realistic existence that takes up space and time duration, it is not categorically nothing either. This is the reason the *wu* 無 in the passage is translated here into 'Nonbeing' rather than 'Nothing.' It is the nothing that has the tendency of becoming something. In the 1970s, Stephen Hawking, in his singularity theorem, proposed a set of sufficient conditions for the existence of a singularity in space–time that could explain the very early state of the newborn universe. In this theory, he notes that the singularity is the zero-size universe at the zero-time of the Big Bang, a 'point' stage similar to the non-space-time complex of Nonbeing at the Beginning in the *Zhuang Zi* cosmogony. He also predicts with mathematical proof that black holes have such strong gravitational force that 'there must be a singularity of infinite density and space–time curvature within a black hole' at which 'the laws of science and our ability to predict the future would break down.' And the black hole singularity is 'rather like the big bang at the beginning of time.'[10]

Hawking then predicts that at the brim of a black hole, pairs of particles and antiparticles, one with positive energy and the other with negative energy, constantly come into being together and also immediately seek out each other and annihilate each other.[11] His predictions are indeed proved by later observations of black hole emission. The coming into being and the annihilation of these particle and antiparticle pairs happen so fast that one cannot say the particles and antiparticles actually exist; then, on the other hand, one cannot say they are not there or that there is absolutely nothing going on, either. It is a state where things do come into being while nothing actually lasts long enough to be something, it can only be regarded, perhaps, as a state loaded with the tension and a tendency toward becoming something, becoming existence. Since the black hole with a singularity inside is similar to the Big Bang at the beginning of time, such a state of constant, almost simultaneous creation and annihilation of particles and antiparticles is very similar to the third stage of the *Zhuang Zi* cosmogony, where, at the Beginning, there is the Nonbeing that has the tendency of becoming Being, the state that actually contains no space and no duration but cannot be humanly differentiated from the state of Being that solidly takes space and time.

Hawking also states that due to the extremely strong gravitational force of the black hole, some of the particles or antiparticles that have negative energy fall into the black hole, leaving the other half of the pair free to either fall into the black hole or escape it due to its positive energy.[12] When it escapes, it is then solid matter born out of the black hole and will travel across the universe, continuing to exist as a black hole emission. In terms of the *Zhuang Zi* cosmogony, it thus becomes the Being 有. At the brim of the black hole, the birth and annihilation of particle/

10 Stephen Hawking, *A Brief History of Time* (New York, NY: Bantam Books, 1998), 90-1.
11 Ibid., 110-11.
12 Ibid.

antiparticle pairs and the formation of matter from escaped particles or antiparticles happen at the same time; one process cannot be distinguished from the other, as they are both likely. It is thus similar to the description in the *Zhuang Zi* passage that at the Beginning, 'suddenly there is Being and Nonbeing. But I don't know, between Being and Nonbeing, which is actually Being and which is Nonbeing.'

The regressive inquiry of this paper starts from the last stage (Stage Four) of the *Zhuang Zi* cosmogony, which is the cosmos, the space–time complex of the Being 有 in proposition four. It then goes backwards through the third stage, the beginning of the cosmos, the 'point' stage of the non-space-time complex of the Nonbeing 無 and the Beginning 始 in propositions one and five. Let's then move further back to the second stage of the *Zhuang Zi* cosmogony, the state before the cosmos as is described in proposition two, 'there is the not yet beginning to have the Beginning' 未始有始, and proposition six, 'there is the not yet beginning to be Nonbeing' 未始有無. To simplify the expression, the stage before the cosmos can be considered as a unity formed by the Not-yet-Beginning in terms of time and the Not-yet-Nonbeing in terms of space. At this stage, there is not even the tendency for something to start coming into being. It is a state that is self-contained, without even the tendency to change, and thus eludes all differentiation, because there simply is no state other than itself. Since time is a concept that has an ultimate dependence upon differentiation, it is meaningless to talk about time in such a state of absolutely no change, not even the tendency to differentiate and change. This is the stage of the no-space-no-time complex, a state without invoking time, or space.

Does this Not-yet-Beginning and Not-yet-Nonbeing unity, this no-space-no-time complex, have a boundary? What is before it? The regressive inquiry then reaches the very first stage of the *Zhuang Zi* cosmogony as described in proposition three, 'there is the not yet beginning to have that not yet beginning to have the Beginning' 未始夫未始有始, and proposition seven, 'there is the not yet beginning to be that not yet beginning to be Nonbeing' 未始夫未始有無. This stage is a unity of the Not-yet-Not-yet-Beginning in terms of time and the Not-yet-Not-yet-Nonbeing in terms of space, or rather in terms *beyond* time and space. Since both concepts of time and space lose their meaning in the second stage of Not-yet-Beginning and Not-yet-Nonbeing, they cannot have meaning in the first stage, either; otherwise, there would be a change in the transition from first stage to the second, and the second stage would no longer be free of change and differentiation. In that case, the only point of change would have been the actual Beginning. Thus the first stage is in fact the same as the second, free of change and differentiation. If the regressive inquiry keeps going back, by the same logic any stage before the first should also be the same, free of change and differentiation, or otherwise, a change would be present and would render the subsequent stage the actual Beginning. The function of offering the first stage, just the same as the second, in the *Zhuang Zi* cosmogony, is to ingeniously form a recursive line of reasoning. By the logic of such recursive reasoning, we reach the idea that all stages, if there are any further ones, beyond Stage One, described in propositions three and seven, should be the same self-contained stage without change or differentiation. It is a boundary-less realm of no time and no space—an everlasting, constant, unchanging existence before the beginning of the cosmogonic process, and thus can be called the infinity–eternity complex.

Therefore, based on the cosmogony analyzed above, time and space are born with the cosmos. They are the defining measurements of the cosmos's existence, and, in turn, they depend on the reality, the changes and differentiations within the cosmos, to be manifested. The complex of eternity and infinity, however, is beyond the boundaries of time and space, indefinable and unmeasurable in terms of these concepts of differentiation; it is simply beyond the reality of the cosmos. The cosmogonic process in the *Zhuang Zi* passage comes forth from within the infinity–eternity complex in Stage One and the no-space-no-time complex in Stage Two, which by recursively mirroring each other signifies the self-containing and self-sustaining nature of the one and same state. This constant, unchanging state takes on two descriptions: as a state beyond time and space in Stage Two and as a state that is boundary-less in Stage One, which in essence is the same state. At some point, there all of a sudden is the tendency to start to come into being, which is the state of the Beginning and the Nonbeing, the non-space-time complex described in Stage Three. Stage Three takes no space and no time duration; once the tendency is there, from this point on, time, space, the cosmos, and the reality that can be measured by time and space all come into being at once.

The Body and Returning to the Dao

A closer look at the description of the Dao discussed above reveals that both the first and the second stages in the cosmogony depicted by the seven propositions are in fact descriptions of the Dao. In the *Zhuang Zi* chapter entitled 'The Great and Venerable Teacher,' the Dao 'has its reality and its manifestation, but is without action or form.' Since time is a manifestation of action, change, and causality, if there is action, there is sequence and duration; then there is time. The Dao has no action, thus the concept of time is meaningless to the presence of the Dao. Form is defined only by a limited amount of space; only when there is extra space outside of the entity has the entity a certain form. The Dao is infinity, thus the concept of space is meaningless to the presence of the Dao. As shown by the four-stage cosmogonic process, the infinity–eternity complex is a state that has no change and no differentiation, and thus has no action. It is beyond the description and measurements of space, and thus has no form. Agreeing with each other, the description in 'The Great and Venerable Teacher' is a direct description of the Dao's attributes, while the analysis of the seven propositions presents the Dao from the existential perspective and reveals it as the presence beyond time and space. Moreover, in 'The Great and Venerable Teacher,' the Dao 'is its own source, its own root,' and 'even before heaven and earth were formed, it has been the absolute existence since antiquity.' It is the infinity–eternity complex that is its own cause as identified through the analysis of the cosmogonic process and shown by the recursive regressive reasoning that there is no prior state of presence that can be different from it.

The Dao, at the point of the Beginning and the Nonbeing (Stage Three), starts to cause the cosmos to come into being. The cosmos is then formed together with its defining dimensions of time and space; it is the Being (Stage Four). In 'The Great and Venerable Teacher,' the formation of the cosmos is described as the Dao 'gave spirituality to the spirits and gods' and 'gave birth to heaven and to earth.' But the Dao, the infinity–eternity complex, should still be present together with the cosmos by its very definition of eternity and infinity and by its own unchanging, self-sustaining nature. Indeed, Guo Xiang's commentary states that 'there is no place where the Dao is not. Therefore, being high it is of no height; being long-lasting it is of no duration; being deep it is of no depth; being old it is of no age; there is no place where it is not, while everywhere it is it is not there.' It is present now and everywhere, but cannot be detected by measuring standards such as height, depth, ancientness, and duration in the reference frame of the cosmic time and space. From the perspective of the cosmos, it is omnipresent and is present everywhere throughout the cosmos, and yet its presence cannot be pin-pointed to anywhere spatially at all, because it is beyond the reality of the cosmos and the concept and measurements of space; it is eternal, lasting forever and at the same time not lasting at all, because in its presence the concept of time does not apply; it does not last, it simply is. It is an all-inclusive existence that is all-of-itself and all-at-once. It is always present and omnipresent, a presence that is infinite and eternal, beyond the reality of cosmic time and space. It cannot be reached from within the confinement of the cosmic time and space.

Thus, from this perspective, some attributes of the Dao are similar to those of a creator, as the Dao 'gave birth to heaven and to earth.' For human beings, who are confined within the time and space of the cosmos, the Dao 'can be transmitted but cannot be owned; it can be understood but cannot be seen.' It cannot be seen, cannot be possessed, because it is well beyond our limited reality, which actually is created by and from it. What is created cannot own its creator, nor can it face the reality of its creator. Interestingly, in the *Zhuang Zi*, it seems unimportant where or how the tendency for the cosmos to come into being actually comes forth. The Daoist philosophy and worldview focus instead on returning to the origin, to the Dao. Different from Hawking's claim that any information before the Big Bang is beyond the current time frame and thus cannot be accessed, the *Zhuang Zi* states that the Dao can be transmitted and understood. For Hawking, it is not possible for human beings to understand the state before the formation of our universe. But in the *Zhuang Zi*, the state before the cosmos is the Dao, and human beings can receive information from it and understand it. Moreover, the space–time boundaries that confine us within our cosmos can be transcended through understanding the Dao, returning to it, and becoming one with it.

How does one transcend the boundaries of one's own existence, rise above the reality of this cosmos defined by time and space, and enter the boundary-less realm of the Dao? First, let's look at a passage in 'Zhi Travels North' 知北遊, the last of the Outer Chapters of the *Zhuang Zi*.

> 冉求問於仲尼曰，未有天地可知邪？仲尼曰，可。古猶今也。
>
> Ran Qiu asks Confucius, 'Can we know about the state when there is not yet heaven and earth?' Confucius replied, 'We can. Antiquity is just like the present.'

This is the imagined character of Confucius speaking here. The *Zhuang Zi* often invokes the characters of Confucius and his disciples and employs them in discussions and parables that expound and confirm Daoist ideas and principles. Here the Confucius character confirms that the Dao, which is the only presence before the cosmos, can be understood. The comment 'antiquity is just like the present,' taken as a comment on the Dao, states the constant, unchanging nature of the Dao. When Ran Qiu gets confused and asks the same question again the next day, 'Confucius' comments further:

> 無古無今，無始無終。未有子孫而有子孫可乎？
>
> There is no antiquity and no present; no beginning and no end. One has yet to have descendants and yet there are descendants, can this be?

The Dao is eternal, forever-lasting, and at the same time all-at-once. To the Dao, there is no difference between antiquity and the present, between beginning and end. The differentiation of these concepts is only an idea within the limited existence of the cosmos that is governed by time and space. To the Dao, all time is present time, or rather, there is no time. Thus, a person's life from the Dao's perspective is all present time, all happening at once. One can have descendants and not have descendants at the same time; there is no difference. All differentiations and changes of stages during a man's life, and even during that of the cosmos itself, are reduced to nothing and zero time in the face of the eternal presence of the Dao. In fact, in the timeless presence of the Dao, time itself is reduced to nothing.

'Confucius' then moves on to discuss the order of life and death from the same perspective:

> 不以生生死，不以死死生。死生有待邪？皆有所一體。有先天地生者物邪？物物者非物，物出不得先物也，猶其有物也。猶其有物也，無已！
>
> Life is not what brings forth death; death is not what ends life. Is there an order of death and life? They both belong to one body [of man]. Are there things that come into being before heaven and earth? What creates things is not a thing; the origination of things does not have a thing prior to it, just as if there were a thing. As if there were a thing, [in fact] there is nothing![13]

Compared to the eternal presence of the Dao, the duration of human life is as short as zero time, and coming into life and dying out are simultaneous, all happening at once, too fast to be differentiated. Therefore, they are but two sides of the very same state that is understood as 'existence,' and it is not one that causes the other. There is no space or distance or time-lapse between these two sides at all, so that they appear to be just the same thing. Thus 'Confucius' asks, 'is there an order of death and life?' Not only is there no sequential order of the two, in the worldview of the *Zhuang Zi* text, they are also treated not in terms of time, but in terms of space, as 'Confucius' adds, 'they both belong to one body.' This explicitly reveals the idea that time and space are always related, if not undifferentiated, concepts in ancient Chinese thought. Normally, when one beholds the body of a human being, all sections of the body—the head, the limbs, the trunk—are present in front of one's eyes. Time, however, confines one to a certain moment, and one cannot behold one's past, present, and future all at once. But from the perspective of the Dao, which is all-of-itself and all-at-once, beholding time is also like beholding a body of space—all the time of the cosmos, the past, present, and future is 'present' in front of the existence of the Dao, as small as a dot. Life and death, as if of the same body, no one is before the other. From this, 'Confucius' thus comments on the origin of the cosmos and states that things cannot originate from another thing, as it merely pushes the real origin further back. 'What creates things is not a thing'; the true origin, thus, must be one that is beyond the reality of the existence of all things. It is the Dao. The worldview of regarding life and death as the same, perhaps, is the reason that the end of the cosmos is not even mentioned in the *Zhuang Zi*. Unlike Hawking and the many theories of modern physics that predict and try to study the end of the universe, in the *Zhuang Zi*, coming into being and dying out are all the same, all part of the nature of a limited existence and all nothing compared to the presence of the Dao. The *Zhuang Zi*'s concern is never the end, but rather how to get back to the origin, to the Dao.

If the lifespan of a man can be viewed in the same way his body of limited space is viewed, the limitations of life and death can then be transcended, as the confinement of the physical body can be transcended. In this way, the mortal being can then reach the boundary-less realm of the Dao. This idea is apparent in the following passages from 'The Great and Venerable Teacher:'

[13] Translations of passages in 'Zhi Travels North' are mine, with reference to Wang Xianqian, *Zhuang Zi jijie*, 144.

> 子祀子輿子犁子來，四人相與語曰，孰能以無爲首，以生爲脊，以死爲尻？孰知生死存亡之一體者，吾與之友矣。四人相視而笑，莫逆於心，遂相與爲友。
>
> Master Si, Master Yu, Master Li, and Master Lai, the four of them were talking together and said, 'Who can regard the Nonbeing as the head, life as the back, and death as the buttocks, who knows that life and death, existence and annihilation, are of one body? I will be his friend!' The four men looked at each other and smiled. No disagreement in their hearts, they then became friends with each other.[14]

Here, the four friends all regard highly the person that 'can regard the Nonbeing as the head, life as the back, and death as the buttocks.' This is a worldview rooted deeply in the *Zhuang Zi* cosmogony, where the cosmos begins with the Nonbeing at the Beginning. It is reflected in the contemplations on life that life, too, like the cosmos, begins with the Nonbeing. From the perspective of the Dao, viewing from the very origin of the cosmos beyond time and space, one sees 'that life and death, existence and annihilation, are of one body.' The next passage builds on the analogy of the body discussed above:

> 子桑戶孟子反子琴張，三人相與友。曰，孰能相與於無相與，相爲於無相爲？孰能登天遊霧，撓挑無極，相忘以生，無所終窮？三人相視而笑，莫逆於心，遂相與友。
>
> Master Sanghu, Meng Zifan, and Master Qinzhang, the three of them were friends with each other and said, 'Who can get along with others without [intentionally] getting along with others? Who can work with others without [intentionally] working with others? Who can climb up to heaven and wander in the mists, roam in infinity, and live without end by means of mutual obliteration?' The three men looked at each other and smiled. No disagreement in their hearts, they then became friends with each other.[15]

Continuing the analogy of the body, this passage praises the person who works with others in the same way the parts of the body work naturally with one another without intentional effort. If all things become one within the Dao, then naturally they work and get along with one another without effort. Such a person can also 'climb up to heaven and wander in the mists, roam in infinity, and live without end by means of mutual obliteration.' This is the definitive, conventional image of a person who has transcended the limitations of his own mundane existence and become one with the Dao in the boundary-less realm of infinity and eternity. It is a realm of total freedom, where one forgets all constraints by and connections with other things, because there is no differentiation between oneself and others and the Dao itself when one is inside the undifferentiated realm of the Dao. This also ties back to Guo Xiang's interpretation of the seven propositions that identify the 'obliteration' 忘 as the way to transcend the limitations of the self and the knowledge-consciousness.

With life and death treated as parts of the body itself, and all parts expected to be undifferentiated and working naturally together within the Dao, transcending the limited existence of a human being and entering the realm of the Dao then depends on how one regulates one's physical body. In the *Zhuang Zi*, many passages are devoted to the regulation of the body and the transcendence of the physical form. Since the Dao is beyond the reality of the cosmos, to return to it, one, as a being confined by time and space, has to start with something tangible, something

[14] Based on *Chuang Tzu: Basic Writings*, trans. Burton Watson, 80; Wang Xianqian, *Zhuang Zi jijie*, 42.
[15] Based on *Chuang Tzu: Basic Writings*, trans. Burton Watson, 82; Wang Xianqian, *Zhuang Zi jijie*, 43.

within the physical cosmos one can reach and work with. Then the body, the embodiment of existence in terms of space and time, becomes the natural start of the way leading to the Dao. As one example among many, a passage in 'Tian Zifang' 田子方, the second to last of the Outer Chapters of the *Zhuang Zi*, makes a connection between the death of the body and the returning to the Dao:

> 孔子見老聃，老聃新沐，方將被髮而乾，慹然似非人。孔子便而待之，少焉見曰，丘也眩與？其信然與？向者先生形體掘若槁木，似遺物離人而立於獨也。老聃曰，吾遊心於物之初。
>
> Master Kong (i.e. the Confucius character in the *Zhuang Zi*) went to see Old Master Dan. Old Master Dan had just bathed and spread his hair over his shoulders to let it dry. He was motionless, and did not seem to be human. Stepping aside, Master Kong waited for a short while before presenting himself and saying, 'Was Kong Qiu dizzy [and seeing things]? or was it real? Just now, Sir, your form and body were stiff like a dried-up tree. You seemed to have abandoned all things, taken leave of men, and been standing in solitude by yourself!' Old Master Dan said, 'I had sent my mind on a journey to the Beginning of things.'[16]

Old Master Dan—that is, Lao zi 老子 (ca. 6th century B.C.)—is the character the *Zhuang Zi* text makes to invoke the founding father of Daoist thought and philosophy to explain key concepts and ideas of Daoism. The 'dried-up tree' or the 'withered tree' is the image frequently employed in the *Zhuang Zi* to describe the body of a person who has transcended the limited existence within the cosmos, returned to the origin, and become one with the Dao. Here, by the 'Beginning of things,' Lao Dan is referring to the Dao. Once one is with the Dao, from the perspective of infinity and eternity, it is meaningless to differentiate one moment from another, one body of space from another—all is in one, which is the Dao, and all is at once and eternal together with the Dao. Thus all things of the cosmos are undifferentiated, equalized, as in the passage right after the one with the seven propositions in the chapter 'Discussion on Making All Things Equal':

> 天下莫大於秋豪之末，而太山爲小。莫壽於殤子，而彭祖爲夭。天地與我並生，萬物與我爲一。
>
> In the world there is nothing bigger than the tip of an autumn hair, and Mount Tai is deemed small; no one has lived longer than a child who died young, and Pengzu is deemed short-lived. Heaven and earth exist with me in parallel, and the ten thousand things are one with me.[17]

With infinity and eternity as reference, the differences between the multitudes of phenomena in the cosmos are infinitely close to zero, thus all things can be equalized. The long life of Pengzu, the one who enjoyed extraordinary longevity and the short life of a child who died young are both within the finite boundaries of time, and both nothing compared to eternity; the biggest mountain and the tip of a hair are both within the finite boundaries of space, and both nothing compared to infinity.

16 Wang Xianqian, *Zhuang Zi jijie*, 43.

17 Translation based on *Chuang Tzu: Basic Writings*, trans. Burton Watson, 38; Wang Xianqian, *Zhuang Zi jijie*, 13.

On an epistemological level, the 'ten thousand things' to be taken as equal also include human judgments, knowledge, and understandings of the ten thousand things within the cosmos. In 'Autumn Floods' 秋水, the tenth of the Outer Chapters of the *Zhuang Zi*, the Lord of the River puts questions to Ruo of the North Sea about judgments between what is noble and what is lowly. Ruo of the North Sea answers that 'viewing things by [the principles of] the Dao, there is nothing noble or lowly' 以道觀之，物無貴賤, and later 'viewing things by [the principles of] the Dao, such ideas of what is noble and what is lowly are merely expressions of superfluous contrasts' 以道觀之，何貴何賤，是謂反衍.[18] Knowledge and understandings, too, should be equalized. In the preamble to the seven propositions in the chapter 'Discussion on Making All Things Equal,' once a statement is made, no matter what it is about, it falls into the same category of ideas that are formulated and expressed in language. Therefore, 'belonging or not, combined together they become one category, then it has nothing to be differentiated from others' 類與不類，相與爲類，則與彼無以異矣. From this perspective, the understandings of the origin of the cosmos in modern physics, in the chapters of the *Zhuang Zi*, and even in the various religious beliefs of mankind, all belong to one large category that is expressed through, and at the same time confined by, language, which becomes nothing when compared to the inexplicable truth of the Dao. No wonder Roger Ames in his *The Art of Rulership* claims that the seven propositions in the *Zhuang Zi* passage are in fact 'an illustration of the imprecision of language.'[19]

Conclusion

The seven propositions in the 'Discussion on Making All Things Equal' chapter of the *Zhuang Zi* are treated here as a regressive description of the cosmogonic process from the perspective of the intertwined relationship of time and space. Such an analysis reveals a cosmogonic process that can be related and compared to that of the Big Bang Theory in modern physics. In so doing, this paper provides an interpretation of the Dao, the origin of the cosmos in the *Zhuang Zi* text, as the infinity–eternity complex that is beyond the concepts of time and space. The boundary-less realm of the Dao is free of any spatial and temporal restraints, free of any kind of differentiation. For human beings, to transcend the confinement of the physical and temporal dimensions of the cosmos is to return to the origin and become one with the Dao. Treating time in terms of space, and the lifespan of man in terms of the human body itself, the *Zhuang Zi* offers a way to transcend the boundaries of space and time, life and death, through transcending the physical existence of the body. Once one escapes the reality of the physical world and becomes one with the Dao in infinity and eternity, there is no need to differentiate the ten thousand things and phenomena of the world. Compared to infinity and eternity, they all become equal in their infinitesimal existence. Thus, the *Zhuang Zi*'s worldview that all things should be deemed equal is a natural result derived from its cosmogony, which takes its origin from beyond the concepts of time and space.

---

18 Translation modified from *Chuang Tzu: Basic Writings*, trans. Burton Watson, 102; Wang Xianqian, *Zhuang Zi jijie*, 43.
19 Roger Ames, *The Art of Rulership* (Honolulu: University of Hawaii Press, 1983), 215.

Measuring Time

The Gifts of Mihragān: Muslim Governers and Gift Giving during Non-Muslim Holidays[1]

Robert Haug

In celebration of the centennial of the Midwest Branch of the American Oriental Society, an organization whose mission and academic interests cross many chronological and geographic divides, the following paper explores the evolution of gift giving during the Zoroastrian celebration of the autumn equinox known as Mihragān (Mihrajān in Arabic) across the historical divide of the Arab–Muslim conquests. Mihragān was a time during which the courtiers, clients, and relatives of kings gave gifts to their rulers. This article examines the ways this holiday was adopted by the Muslim rulers who conquered the Sasanian Empire and governed a large Zoroastrian population—or at least a population whose political culture was heavily influenced by Zoroastrian practice—as well as the challenges posed to Muslim rulers when faced with a non-Muslim holiday that could provide economic and political benefit to themselves and their supporters. In these exchanges we not only see a meeting of varied religious and cultural traditions, but even competing calendar systems and cycles of holidays. In this article I argue that Mihragān formed a 'middle ground' between the Arab-Muslim conquers and their Iranian-Zoroastrian subjects, following the theory first proposed by Richard White in his study of European-Indian interactions in the Great Lakes region of North America.

In the pre-Islamic period, the Zoroastrian celebration of the autumn equinox known as Mihragān was a time when courtiers, clients, and relatives of Iranian kings—including but not limited to the Sasanian emperors—gave their ruler gifts and he, in turn, held a public audience, reconfirmed bonds, and assigned duties. We may assume that the Arab–Muslim conquests of the Sasanian Empire in the first century AH/seventh century CE would have disrupted these traditions by replacing a Zoroastrian ruling class with a Muslim ruling class, but they did not. Instead, the gifts of Mihragān developed into a form of extra-legal taxation in the lands of the former Sasanian Empire. This exchange of gifts was not entirely financial and the gifts traditionally given to Iranian kings held symbolic meaning. Furthermore, the gifts often included items that were discouraged if not outright forbidden by Muslim law and norms, which created a dissonance for Muslim rulers collecting gifts in honor of a Zoroastrian holiday. This article will examine the exchange of gifts between presumably non-Muslim subjects and Muslim rulers during the festival of Mihragān with special emphasis on practices associated with the frontier city of Balkh in Ṭukhāristān (Bactria or northern Afghanistan) during the first century of Muslim rule in the region. It will be argued here that as Muslim governors integrated themselves into the local populations, Mihragān became a site of 'middle ground,' not merely a position of compromise or accommodation between conquerors and conquered but rather 'a set of practices, rituals, offices, and beliefs that although comprised of elements of the group in contact is as a whole separate from the practices and beliefs of all of those groups.'[2]

1 An early version of this paper entitled 'How to be an Arab *Dihqan*' was presented at the Annual Meeting of the American Oriental Society in Phoenix, AZ March 14–17, 2014. The author would like to thank everyone who attended the panel and commented on this earlier draft.

2 Richard White, *The Middle Ground: Indians, Empires, and Republics in the Great Lakes Region, 1650-1815*, Twentieth Anniversary Edition (Cambridge: Cambridge University Press, 2011), p. xiii.

While the celebration of Mihragān as practiced in Iraq and at the court of the caliphs acted as a continuation of royal practices across the conquest divide, in distant sites like Balkh, the celebration of Mihragān played a more important role by kneading together local and imperial elites.

The Festival of Mihragān

Mihragān is a Zoroastrian holiday—though its origins may be older—celebrating the autumn equinox. Traditionally it was related to the Indo-Iranian god Mithra with the name of the holiday itself meaning something like 'festival of Mithra.'[3] Mihragān was held on the day Mihr (the 16th day) of the month Mihr (the seventh month) though the precise date changed over time and the festivities spread out to cover as many as ten days at its lengthiest.[4] Over time, the attachment to Mithra was lost and/or confused and the holiday became more closely associated with the autumn equinox alone and served as a harvest festival.[5] There may be a connection between Mithra's role as an arbiter and his connection to pacts, contracts, and covenants and the equinox being 'the astronomical phenomenon most easily linked to the concept of equity and equilibrium,' but it is unclear whether such a connection was purposeful or happenstance.[6] Similarly, at least one tradition recorded on the authority of the ninth century Persian philosopher Abū al-ʿAbbās al-Īrānshahrī claims that Nowrūz—the Persian New Year celebrated during the spring equinox and, therefore, Mihragān's mirrored twin—and Mihragān were the days on which God made a treaty between the Light and Darkness or between Ahura Mazda and Ahriman.[7]

Our best source for the pre-Islamic tradition is found in the writings of the Khwarāzmian polymath Abū Rayḥān Muḥammad b. Aḥmad al-Bīrūnī (362–440/973–1048). In his *al-Athār al-baqqiya ʿan al-qurūn al-khāliya* (*The Vestiges of Past Centuries*, better known as *The Chronology of Ancient Nations*), Bīrūnī gives a number of meanings for the name of the festival that span from the love (*mihr*) of the spirit/self (*jān*)—an etymology popular with Persian poets[8]—to a celebration of the sun (also *mihr*). According to proponents of this latter tradition, the sun first appeared on Mihragān and this connection was acknowledged by the Sasanian emperors who wore a crown with an image of the sun and the wheel on which it turns during Mihragān festivities.[9] According to the historian and geographer Abū al-Ḥassan ʿAlī b. Ḥusayn al-Masʿūdī (282–345/896–959), Mihragān was the first day of winter, but Bīrūnī argues that this ceased to be the case as people neglected intercalation—the insertion of extra days into the calendar to maintain synchronicity with the seasons.[10] In the *Kitāb al-tāj fī akhlāq al-mulūk* (*The Book of the Crown in the Ethics of Kings*) of Muḥammad b. al-Ḥārith al-Taghlibī or Thaʿlabī—traditionally attributed to his much more famous contemporary, the Iraqi litterateur Abū ʿUthmān ʿAmr

3 On the etymology, Jean Calmard, 'Mihragān,' *Encyclopaedia of Islam2* vol. 7 (Leiden: Brill, 1990), 15.

4 For a detailed study of the place of Mithra in the Iranian calendar, Walter Belardi, *Studi mitraici e mazdei* (Rome: Istituto di glottologia della Università: Centro culturale italo-iraniano, 1977), 61–149.

5 It should be noted that this was not always the case and under the Arsacid Parthians. Nowrūz, the New Year celebration, and Mihragān swapped positions in the calendar. Calmard, 'Mihragān,' 15. There is some residual confusion on this point in later sources.

6 Simone Cristoforetti, 'Mehragān,' *Encyclopædia Iranica*, online edition, 2013, http://www.iranicaonline.org/articles/mehragan.

7 Abū Rayḥān Muḥammad b. Aḥmad Bīrūnī, *The Chronology of Ancient Nations* (trans. C. Edward Sachau; London: Oriental Translation Fund, 1879), 208.

8 Calmard, 'Mihragān,' 15.

9 Bīrūnī, *The Chronology of Ancient Nations*, 207.

10 Abū al-Ḥasan ʿAlī b. al-Ḥusayn Masʿūdī, *Les prairies d'or* (trans. C. Barbier de Meynard and A. Pavet de Courteille; Paris: Imprimerie Nationale, 1861), vol. 3, p. 404; Bīrūnī, *The Chronology of Ancient Nations*, 209.

b. Baḥr al-Jāḥiẓ (160–255/776–868-9)—Mihragān ushered in the winter rains and cold just as Nowrūz brought heat.[11]

For the celebration of Mihragān in both the Sasanian and early Islamic periods, we must rely on sources in Arabic primarily from the fourth/tenth century and later. In these sources, gift giving plays a regular part, with traditions of gift giving perhaps going back to the Arsacid Parthians (247 BCE–224 CE).[12] In *al-Maḥāsin wa'l-aḍdād* (*The Beauties and their Opposite*) a 'mirror for princes' also erroneously attributed to Jāḥiẓ, these gifts appear extremely ceremonial. A silver table (*khuwān*) would be brought before the king on which was placed seven flat breads (*arghifa*) prepared with seven different grains (the text specifies wheat, barley, pearl millet, durra or sorghum, chickpeas, lentils, rice, sesame, and different types of beans (*al-bāqillayī wa'l-lūbiyā*)), seven branches of an auspicious tree (the text specifies olive, quince, and pomegranate), seven white bowls (*sukrujāt*), seven silver dirhams minted that year, a new gold dinar, and a bouquet of esfand or wild rue.[13] The author of *al-Maḥāsin wa'l-aḍdād* continues to say that the first thing the king would be given was a serving tray of gold or silver (*ṣīnīya thahab aw fiḍḍa*) adorned with white sugar, shelled Indian walnuts (candlenuts?) (*jawz hindī muqashshar raṭb*), and silver or gold cups containing dates soaking in cow's milk (*jāmāt fiḍḍa aw thahab wa yabtada'a bi'l-laban al-ḥalīb al-ṭarīy min-hi*).[14] According to Taghlibī in *Kitāb al-tāj*, the gifts presented to the king were less formal with each man giving what he loved best or what he excelled at producing. Examples given by Taghlibī include musk, ambergris (*ʿanbar*), clothing and robes (*kuswa wa thiyāb*), horses, lances (*rumḥ*), swords, arrows, gold, and silver.[15] Poets gave poems while scribes gave speeches and boon companions (*nadīm*) gave gifts of novelties or art objects (*al-tuḥafa wa al-ṭurfa wa al-bāk*) and fresh produce (*al-bākūra min al-khaḍrāwāt*).[16] The value of these gifts was recorded in the *dīwān* and the king remunerated them with his own gifts and rewards making this a reciprocal exchange. Beyond the exchange of gifts, Mihragān was an occasion for the king to renew the bonds with his subjects and he would assign duties to the people, presumably the nobles and courtiers who joined him for the festivities, tying the celebration to the themes of pacts, contracts, and covenants mentioned above.[17]

In as much as Mihragān was a celebration of the changing seasons, the exchange of clothing was also highlighted. According to Taghlibī, Mihragān was when the king would dispense of his summer clothes for his winter clothes and on the day of Mihragān itself he would dress in a new outfit of poplin (*khazz*), embroidered silk, or *mulḥam* (a combination of cotton and silk woven together).[18] Such traditions continued into the Islamic period, appearing perhaps as early as the reign of the Caliph Muʿāwiya (r. 41–60/661–680). The Andalusian poet and *belle-lettrist* Aḥmad b. Muḥammad b. ʿAbd Rabbih (246–328/860–940) reports in his *al-ʿIqd al-farīd* (*The Unique Necklace*) that the governor of Iraq Ziyād b. Abīhi (d. 53/673) commanded the chamberlain to not admit anyone wearing unseasonable clothing, but in this context people who wear winter clothing

11 Pseudo-Jāḥiẓ, *Kitāb al-tāj fī akhlāq al-mulūk*, ed. Aḥmad Zakī (Cairo: al-Maṭbaʿa al-amīriyya, 1914), 146. On the authorship of *Kitāb al-tāj*, see G. Schoeler, 'Verfasser und titel des dem Ǧāḥiẓ zugeschriebenen sog. *Kitāb al-tāǧ*',' *Zeitschrift der Deutschen Morgenländischen Gesellschaft* 130 (1980): 217–25.

12 The reference to the Arsacids giving gifts on Mihrajān can be found in Abū Manṣūr ʿAbd al-Malik Thaʿālibī, *Ghurar akhbār mulūk al-furs wa sīyarihim* (ed. Hermann Zotenberg; Paris: Imprimerie Nationale, 1900), 47. It should probably be treated as a bit of wisdom literature rather than a historical account. See Cristoforetti, 'Mehragān.'

13 Pseudo-Jāḥiẓ, *al-Maḥāsin wa'l-aḍdād* (ed. Gerlof Van Vloten; Leiden: Brill, 1898), 361.

14 Ibid., 362.

15 Pseudo-Jāḥiẓ, *Kitāb al-tāj*, 146–7.

16 Ibid., 148.

17 Ibid., 144.

18 Ibid., 149–50.

in summer and summer clothing in winter are described as the people for whom God does not care.[19] Michael Morony has suggested that this was tied to the traditions of exchanging clothes at Nowrūz and Mihragān though Ibn ʿAbd Rabbih does not mention either holiday in this context.[20] More direct evidence comes after the reemergence of Iranian dynasties with the Ṭāhirid governor of Khurāsān ʿAbdullāh b. Ṭāhir (r. 213–230/828–845) distributing clothing on Mihragān.[21] According to Bīrūnī, the lords of Khurāsān continued until his day (early fifth/11th century) to distribute autumn and winter clothing to their soldiers on Mihragān.[22] With these traditions of Mihragān in mind, let us now turn to one of the earliest examples of a Muslim governor engaging the gifts of Mihragān.

Mihragān, the Conquest of Balkh, and Taxation

In the summer of 33/653, the armies of the Islamic caliphate made their first incursion into Ṭukhāristān led by the general Aḥnaf b. Qays (d. 67/686). They besieged the city of Balkh and obtained a peace agreement from the people according to which the inhabitants of Balkh would pay the Muslims a tribute of 400,000 silver dirhams. From the account of this campaign preserved in the chronicle of Abū Jaʿfar Muḥammad al-Ṭabarī (224–310/839–923) on the authority of Abū al-Ḥasan ʿAlī al-Madā'inī (d. 231/845–6), the author of a lost history of the conquest of Khurāsān, Aḥnaf did not tarry long in Balkh and before the tribute could be collected he had gathered the bulk of his troops and marched for Khwarāzm on the southern shores of the Aral Sea where they campaigned until the onset of the winter cold forced the Arabs to turn back south. While Aḥnaf was leading his troops approximately 1,000 km away, he left his nephew Asīd b. al-Mutashammis at Balkh to supervise the collection of the tribute.[23] According to the account preserved by Ṭabarī, one day the people of Balkh came to Asīd bringing him gifts (*hādāyā*) of gold and silver vessels (*āniya*), gold and silver coins (*danānīr wa darāhim*), furniture or household goods (*matāʿ*), and robes (*thiyāb*). Asīd asked if these gifts were a part of the tribute and he was told no, 'this is rather something we do on this day for our ruler in order to conciliate him.' Asīd inquired about the day and was told it was Mihragān, to which he replied, 'I do not know what that is.'[24]

Asīd's response to the gifts of Mihragān is at least in part a literary device that highlights a common theme in narratives of conquest, the conqueror's ignorance of and subsequent response to the cultural norms and practices of the people they have conquered. While these interactions may often be discussed in terms of accommodation, identifying what the conquerors borrowed from the conquered and what the conquered borrowed—or had foisted upon them—from the conquerors, such exchanges can be much more complex and nuanced. In his study of European expansion into the Great Lakes region of North America and their interaction with indigenous populations, Richard White proposed the concept of the 'middle ground' which he has defined as:

19 Aḥmad b. Muḥammad b. ʿAbd Rabbih, *al-ʿIqd al-farīd* (Beirut: Dār al-kutub al-ʿilmiyya, 1983), vol. 1, p. 64.
20 Michael G. Morony, *Iraq after the Muslim Conquest* (Princeton, NJ: Princeton University Press, 1984), 73.
21 Pseudo-Jāḥiẓ, *Kitāb al-tāj*, 149.
22 Bīrūnī, *The Chronology of Ancient Nations*, 209.
23 This portion of the account is confirmed in Aḥmad b. Yaḥyā Balādhūrī, *Kitāb futūḥ al-buldān* (ed. Michiel Jan de Goeje; Leiden: Brill, 1866), 408.
24 Abū Jaʿfar Muḥammad b. Jarīr Ṭabarī, *Ta'rīkh al-rusul wa'l-mulūk* (ed. Michiel Jan de Goeje; Leiden: Brill, 1879), vol. 1, p. 2903; Abū Jaʿfar Muḥammad b. Jarīr Ṭabarī, *The History of Ṭabarī* 15 (trans. R. Stephen Humphreys; Albany, NY: State University of New York Press, 1987), 106–7.

> [an] attempt to understand the world and the reasoning of others and to assimilate enough of that reasoning to put it to their own purposes... Perhaps the central and defining aspect of the middle ground was the willingness of those who created it to justify their own actions in terms of what they perceived to be their partner's cultural premises. Those operating in the middle ground acted for interests derived from their own culture, but they had to convince people of another culture that some mutual action was fair and legitimate... The congruences arrived at often seemed—and, indeed, were—results of misunderstanding or accidents.[25]

Asīd's response to the gifts of Mihragān seems to come directly from such a place of ignorance and misunderstanding. Not only does he respond to the gifts and the initial explanation of them with a clear 'I do not know what that is,' but, as we will shortly see, in his attempt to decipher the meaning of the gifts and, more importantly, their licitness he is more focused on understanding the young Islamic empire whose own cultural and political norms were still forming a short 20 years after the death of the Prophet Muhammad rather than the cultural norms of the people of Balkh.

Following Asīd's declaration that he did not know what Mihragān was, Madā'inī's account—as preserved by Ṭabarī—shifts the focus to the proper ownership of the gifts from the perspective of the caliphate. Asīd was concerned whether or not the gifts were rightfully his and put them aside to await the advice of his uncle. His concerns seem to rest primarily on the question of proper ownership and not on understanding the culture or motivations of the people of Balkh. Could he, as the lieutenant of his uncle, keep valuable gifts presented by recently conquered peoples? Should these gifts be considered part of the tribute and, if so, would they therefore need to be distributed according to the laws governing taxation, tribute, and spoils among the warriors who conquered Balkh, their commander, the caliph, and, ultimately, the Muslim community as a whole?[26] Upon Aḥnaf's return to Balkh, Asīd asked his uncle about the gifts, but Aḥnaf was likewise uncertain of their status and decided to take them to Baṣra where he could consult his overlord, the governor ʿAbdullāh b. ʿĀmir b. Kurayz (d. ca. 57–9/ca. 678–80) who oversaw Khurāsān and the east from Iraq. Ibn ʿĀmir ultimately ruled that Aḥnaf could keep the gifts for himself, but, in what may be a show of pious humility, Aḥnaf declared that he had no need for them and so Ibn ʿĀmir instead gave the gifts of Mihragān to a man named Mismār al-Qurashī who 'seized them greedily.'[27] Poor Asīd was never given the opportunity to claim for himself the gifts that were first presented to him.

The way Ṭabarī presents this anecdote emphasizes the economic aspects of this exchange. Asīd was awaiting the collection of the tribute and, when presented with the gifts, he initially asked if they belonged to the tribute. The questions posed to Aḥnaf and Ibn ʿĀmir then relate to proper ownership of the gifts of Mihragān, not whether a Muslim governor could participate in a non-Muslim holiday, and the whole story resolved once the gifts found an owner in Mismār, confirming that Arab–Muslims could accept the gifts of Mihragān. Modern scholars have likewise tended to focus on the fiscal aspects of the gifts of Mihragān with similar gifts given during the spring equinox, on the occasion of the Persian New Year or Nowrūz. As Simone Cristoforetti summarized in the *Encylopædia Iranica*, 'The term 'gifts of Nowruz and Mehrajān'

25 White, *The Middle Ground*, 52.

26 For an overview of the rules for distributing spoils, see Frede Løkkegaard, *Islamic Taxation in the Classic Period with Special Reference to Circumstances in Iraq* (Copenhagen: Banner & Korch, 1950), 41–5.

27 Ṭabarī, *Ta'rīkh al-rusul wa'l-mulūk*, vol. 1, pp. 2903–4; Ṭabarī, *The History of al-Ṭabarī* 15, 107.

attested since the early Islamic period should mean a kind of tax apart from those under Sharia law.'[28] We know that already during the Umayyad period, the legality of acceptance of the gifts of Mihragān was debated in terms of taxation. As early as the reign of Muʿāwiya, ʿAbdullāh b. Darrāj, a client or *mawlā* of the caliph entrusted with collecting taxes and maintaining crown lands in Kūfa, increased revenues by 10,000,000 dirhams annually by collecting the gifts of Nowrūz and Mihragān.[29] On the other hand, the Caliph ʿUmar II (r. 99–101/717–20) declared the gifts of Nowrūz and Mihragān illegal in a letter to the governor of Kūfa ʿAbd al-Ḥamīd b. ʿAbd al-Raḥman instructing him to rein in some of the excesses and abuses of his predecessors. In this letter ʿUmar did not focus on the cultural and religious aspects of the gifts of Mihragān but rather classified them among excessive forms of taxation or extortion which were to be curtailed including 'tolls (*āyīn*),[30] fees for mint officials (*ujūr al-ḍarrābīn*), the gifts of Nowrūz and Mihragān (*hadīya al-Nawrūz wa al-Mihrajān*), and fees for official papers (*thaman al-ṣuḥuf*), couriers (*ujūr al-fuyūj*), housing (*ujūr al-buyūt*), and weddings (*darāhim al-nikāḥ*).'[31] Ann Lambton has examined the giving of gifts to rulers in Persianate societies as a form of taxation, including on the occasion of Mihragān, but places its formal development and 'institutionalization' much later, in the Īlkhānid period or the late seventh/13th century.[32] There remains no complete study of the role of the gifts of Mihragān and/or Nowrūz as a form of taxation in the Islamic period.

The conflict between a pre-Islamic and Islamic economy found here has parallels in the 'Markets of the Arabs' traditions examined by Michael Bonner. In the Arabian context, rather than a tradition of courtiers exchanging gifts with their lords around holidays, an annual cycle of markets and fairs developed, also tied to the solar calendar, in which local lords and nobles employed a complex and often opaque system of competitions and displays of generosity to set prices in a manner that reconfirmed their social position and authority. In the Islamic era, these practices were discouraged or outright banned in order to promote a system of prices closer to a modern liberal market economy.[33] In Bonner's analysis, markets and fairs were spaces of competition between pre-Islamic and Islamic economies and gift giving was an important aspect of the pre-Islamic tradition Muslim traditions were hoping to replace. As the gifts of Mihragān were reconfigured as a form of taxation, what Bonner called 'the silent trade of the Arabs' was replaced with a market economy which helped diminish the pre-Islamic economic and political elites.

In the case of the gifts of Mihragān (and Nowrūz), some Muslim rulers may have also used these pre-Islamic traditions themselves as a means to get around Islamic prohibitions and limits on taxation. When it comes to revenue and prestige generation, the non-Islamic origin of the holiday may be its most important feature. As A. Shapur Shahbazi argues, Nowrūz and

28 Cristoforetti, 'Mehragān.'

29 Aḥmad b. Abī Yaʿqūb Yaʿqūbī, *Tārīkh al-Yaʿqūbī* (Beirut: Dār Ṣādr, 1431), vol. 2, p. 218; Muḥammad b. ʿAbdūs Jahshiyārī, *Kitāb al-wuzarā' wa'l-kuttāb*, ed. Muṣṭafā Saqā, Ibrāhīm Abyārī, and ʿAbd al-Ḥafīẓ Shiblī (Cairo, 1938), 24.

30 Arthur Christensen says *āyīn* was the name for the clothing gifted by the king to his courtiers and relatives on Nowrūz and Mihragān. In the context of ʿUmar II's letter to ʿAbd al-Ḥamīd, tolls seems to be the more appropriate translation. Arthur Christensen, *L'Iran sous les Sassanides* (Copenhagen: E. Munksgaard, 1944), 125, 407-415.

31 Ṭabarī, *Ta'rīkh al-rusul wa'l-mulūk* vol. 2, p. 1367; Ṭabarī, *The History of al-Ṭabarī* 24 (trans. David Stephen Powers; Albany, NY: State University of New York Press, 1989), 96. A slightly different version of this letter also appears in Abū ʿUbayd al-Qāsim b. Sallām, *Kitāb al-amwāl* (Cairo: Maktabat al-Kullīyāt al-Azharīya, 1968), 64–6.

32 Ann Lambton, 'Pīshkash: Present or Tribute?,' *Bulletin of the School of Oriental and African Studies* 57 (1994): 145–7.

33 Michael Bonner, 'The Arabian Silent Trade: Profit and Nobility in the 'Markets of the 'Arabs''', in Roxani Eleni Margariti, Adam Sabra, and Petra Sijpesteijn, eds., *Histories of the Middle East: Studies in Middle Eastern Society, Economy and Law in Honor of A.I. Udovitch*, (Leiden: Brill, 2010): 23–51.

Mihragān 'provided the rulers with an alternative source of affirming and enhancing their power and prestige through a strictly non-Islamic channel; for unlike religious festivals, they could appear and be celebrated as the focal point and the peerless heroes of the occasion.'[34] Similarly the non-Islamic character of these holidays may allow rulers to sidestep Muslim restrictions and limitations on taxation, the very situation to which ʿUmar II was responding in the aforementioned letter.

Mihragān and Court Cultures in Transition

The fiscal element and certainly the role of gift giving as a form of taxation is not the only—nor necessarily the central—element of the exchange described by Ṭabarī in the case of Asīd in Balkh though, especially if we would read the anecdote from the perspective of the Balkhīs who, when given the opportunity to include the gifts in the tribute Asīd was waiting to collect, insisted that this was a separate payment to the ruler. While we lack the Balkhīs' voices in our sources, we may challenge Ṭabarī's overt presentation of them as particularly irrational actors who, when burdened with tribute payments to a foreign invader, asked if they could please pay some more taxes? Instead there is a deeper cultural exchange underlying this material exchange between the Balkhīs and Asīd, one that Ṭabarī acknowledges but does not make as prominent a part of the story as Asīd's economic concerns and one which should have raised concerns for Asīd and Aḥnaf as well regarding the legality of the gifts of Mihragān. Perhaps it is an aspect of the exchange we should presume that Ṭabarī's—or earlier Madā'inī's—audience understood without direct prompting.

The gifts given to Asīd b. al-Mutashammis on the occasion of the autumn equinox of 33/653 were not the ceremonial gifts described in *al-Maḥāsin wa'l-aḍdād* and discussed above, but this should not be surprising. The accounts provided in *al-Maḥāsin wa'l-aḍdād*, a 'mirror for princes,' are meant to highlight proper royal behaviors and describe idealized events at the Sasanian court, not the practices one would find in a distant frontier city that only occasionally found its way under the authority of the Sasanian emperors. The gifts in this case are much closer to those described by Taghlibī, but, more importantly, they are focused primarily on items needed for furnishing or decorating the court—gold and silver vessels (*āniya*) and furniture or household goods (*matāʿ*)—and for dressing the ruler—robes (*thiyāb*)—besides the always useful and welcome gold and silver. These are gifts designed to literally dress up Asīd as a ruler on the Iranian model familiar to the Balkhīs. The context around the account as presented by Ṭabarī also ties to the broader theme of preparing the court for winter alongside the Arabs' lack of preparation for the changing seasons on this distant frontier. Ṭabarī emphasizes that Aḥnaf remained in Khwarāzm until winter assailed him (*aqāma ḥattā hajama ʿalī-hi al-shitā'*), forcing him to return to Balkh with the army.[35] This is not the only example of the cold obstructing the conquest of Central Asia—or of Khwarāzm in particular—with the adoption of local clothing being the solution. Fifty years later (85/704), the governor of Khurāsān Yazīd b. al-Muhallab (53–102/673–720) campaigned in Khwarāzm, taking prisoners (*sabīyā*) as part of a peace agreement, but as Yazīd and his men were returning to the provincial capital of Marw the winter cold grew intense and they found themselves unprepared. Yazīd and his men then took

34 A. Shahpur Shahbazi, 'Nowruz ii. In the Islamic Period,' *Encyclopædia Iranica.*, online edition, 2016, http://www.iranicaonline.org/articles/nowrus-ii.

35 Ṭabarī, *Ta'rīkh al-rusul wa'l-mulūk*, vol. 1, p. 2903; Ṭabarī, *The History of al-Ṭabarī* 15, 106.

the prisoners' clothes, thus protecting themselves from the cold but causing their prisoners to die of exposure.[36]

One of the challenges for a Muslim governor receiving the gifts of Mihragān, at least as these accounts detail, is that they consisted largely of objects deemed forbidden to Muslims. Looking only at *Ṣaḥīḥ Bukhārī* for purposes of brevity though similar sentiments may be found in other collections of Prophetic tradition, there are numerous Hadith in which the Prophet Muhammad forbade the use of gold and silver vessels (*āniya*) including injunctions that 'gold, silver, silk, and *dībāj* (silk brocade) are for them [unbelievers] in this world and for you [Muslims] in the hereafter'[37] or 'The Messenger of God forbade us to drink out of gold and silver vessels, or eat from them, and also forbade the wearing of silk and *dībāj* or sitting on it.'[38] In other traditions, still limiting ourselves only to *Ṣaḥīḥ Bukhārī*, the use of gold and silver vessels are listed as one of the seven actions the Prophet Muhammad specifically forbade,[39] but Muslims will eat from silver and gold vessels as a reward in the afterlife.[40] While the accounts provided by Ṭabarī on the gifts of Mihragān focused on the permissibility of accepting the gifts in economic terms, they do not directly address the question of whether or not one would be allowed to actually make use of them as the ceremonial objects they were intended.

We may presume that there were similar problems with the robes gifted to Asīd. Though the material was not specified, it is a safe assumption that robes given to a ruler during the celebration of Mihragān would follow the standards of the Sasanian court and would therefore include a notable component of silk—perhaps in a hybrid cloth such as *mulḥam*—even if they were winter robes and especially in a city such as Balkh so close to the Soghdian trading routes which brought Chinese silk into the Iranian world. Richard Bulliet has fully explored the prohibitions against silk and the resulting development of a cotton industry in early Islamic Khurāsān, which would come long after Asīd's time in Balkh.[41] Like the gold and silver vessels, any silk robes given to Asīd would not be useful to him as robes if he were to maintain even a moderate adherence to Muslim norms. When it comes to such sumptuary laws, as Bulliet further argues, their primary importance may lay in maintaining an outward, physical distinction between the conquerors in cotton and wool and the conquered in silk and brocade at a time when the conquered far outnumbered the conquerors.[42] As such, prohibitions against Muslims wearing silk may act as a kind of reversed Pact of ʿUmar—the document of uncertain provenance which set out laws dictating the behavior of non-Muslim subjects of the caliphate including explicit sumptuary laws.[43] On the other hand, the giving of such gifts could have the opposite effect by transforming a recently arrived foreign conqueror into someone more recognizable to the locals. Such concerns may be even further exacerbated in distant Balkh, a city that would be subjected to a pattern of raids and incursions from not only the Arabs but also Hephthalite and Turkish forces for another century following this initial Muslim occupation.[44]

36 Ṭabarī, *Ta'rīkh al-rusul wa'l-mulūk*, vol. 2, pp. 1142–3; Ṭabarī, *The History of al-Ṭabarī* 23 (trans. Martin Hinds; Albany, NY: State University of New York Press, 1990), 87.

37 *Ṣaḥīḥ al-Bukhārī*, 7:536, 722. A similar sentiment is found in *Ṣaḥīḥ al-Bukhārī*, 7:337, 537.

38 *Ṣaḥīḥ al-Bukhārī*, 7:728.

39 *Ṣaḥīḥ al-Bukhārī*, 2:331, 7:539, 553, 753, 8:253.

40 *Ṣaḥīḥ al-Bukhārī*, 4:468, 469, 544, 6:401, 402, 9:536.

41 Richard Bulliet, *Cotton, Climate, and Camels in Early Islamic Iran: A Moment in World History* (New York: Columbia University Press, 2009), Chapter 2.

42 Ibid. 52–3.

43 For a recent study of the Pact of ʿUmar and the influence of Sassanian sumptuary laws on it, Milka Levy-Rubin, *Non-Muslims in the Early Islamic Empire: From Surrender to Coexistence* (Cambridge: Cambridge University Press, 2011).

44 For an overview of the fate of Balkh in the first/seventh century, see Robert Haug, *The Eastern Frontier: Limits of Empire*

Moving forward several centuries, accounts of the celebration of Mihragān by the Abbasid Caliphs imply that this was a day to transgress Muslim norms and behave as the Sasanian emperors had done before. Due to the context under which most of our sources were written, primarily the fourth/tenth century, evidence from the Abbasid and later periods is much more abundant and therefore provides us with a more detailed dataset. When these sources discuss events during the conquests, their audience included members of the Abbasid court and there is certainly some form of didacticism at play as well and therefore accounts of the conquests and the behaviors of the Abbasid caliphs may be in conversation with each other. Furthermore, the association made between the Abbasid court and the Sasanian court which they are said to have consciously modeled brings further points of comparison to the fore besides the fact that such ostentatious flaunting of religious norms was perhaps more prevalent and permitted at the imperial court. Masᶜūdī gives a report of the Caliph al-Rāḍī (r. 322–329/934–940) celebrating Mihragān at the home of Bajkam (d. 329/941), the Turkish *amīr al-umarā'* or commander of the commanders who had seized effective control of the caliphate in the final years of Rāḍī's reign. According to this report, the caliph's tutor, Abu al-Ḥasan al-ᶜArūḍī (d. 342/953–4), found Rāḍī isolated from the raucous celebration, depressed by Bajkam's grip on power as symbolized by a commemorative coin Bajkam had struck for the occasion. In an attempt to cheer up the caliph and encourage him to join the celebrations, ᶜArūḍī recited some lines of poetry attributed to the Caliph al-Ma'mūn (r. 198–218/813–833) on the occasion of Mihragān.

> Offer the boon companions, on this day of Mihragān,
> lined up around the ancient wine jug
> A cup of the old Iranian royal wine (*khusrawānī*)
> for this feast is the feast of the Khusrows
> And keep away the drinkers of *zabīb* (raisin liquor) without exception
> for the desires of those who drink *zabīb* disagree with my desires.
> I drink it and I believe it is forbidden
> and I ask for permission from the Lord to whom I am indebted
> They drink it and believe it to be permitted
> And that is miserably two sins

Upon hearing these words, Rāḍī decided to join the celebration.[45] Ma'mūn's poem and the context in which ᶜArūḍī deploys it emphasizes the custom of Mihragān to behave as the Sasanian emperors had, despite the fact that such behavior transgressed Muslim norms—the drinking of wine in this case—but that one must seek forgiveness for doing so. Furthermore, one should do so fully in the custom of the Sasanians, enjoying the religiously forbidden *khusrawānī* wine instead of the more popularly accepted *zabīb* raisin liquor.[46]

Making such transgressions was not always as easy as asking for forgiveness after the fact. Abū al-Faḍl Muḥammad b. Ḥusayn Bayhaqī (385–470/995–1077) reports that in 422/1031, the festival of Mihragān overlapped with the Ramaḍān fast, forcing the Ghaznavid Amīr Masᶜūd (r. 421–432/1030–1041)—at whose court Mihragān celebrations were a staple part of the calendar[47]—

in Late Antique and Early Medieval Central Asia (London: I.B. Tauris, 2019), 85–90.

45 Masᶜūdī, *Les prairies d'or*, vol. 8, pp. 340–3.

46 On the discourse around the permissibility of different forms of wine, see Kathryn Kueny, *The Rhetoric of Sobriety: Wine in Early Islam* (Albany, NY: State University of New York Press, 2001). Specifically on the production of *zabīb*, pp. 71–4.

47 Bayhaqī laments at one point that when Masᶜūd's life came to an end so did the days of Mihragān and its celebration. Abū al-Ḥasan 'Alī ibn Zayd Bayhaqī, *Tārīkh-i Bayhaq*, ed. Qāsim Ghanī and 'Alī Akbar Fayyāḍ (Iran: Intishārāt-i Gām va

to accept his gifts on the day of Mihragān but wait two days to enjoy at great expense the accompanying festivities complete with copious amounts of wine, musicians, and poets.[48] A similar problem occurred in 429/1038 when Mihragān fell on the day of ʿArafa (the second day of the Ḥajj). Masʿūd received his gifts on that day but fasted as is customary. Bayhaqī adds that no one feasted or drank that day either in secret (*panhān*) or in public (*āshkārā*), presumably out of fear for Masʿūd, but the following day was ʿEīd al-Aḍḥā, the Feast of the Sacrifice, and Masʿūd organized a feast for his soldiers and courtiers including great amounts of wine in celebration of both Mihragān and ʿEīd al-Aḍḥā. Masʿūd and his boon companions then drank heavily together for a week straight before heading off on campaign.[49] Even among rulers who were usually willing to transgress Muslim norms on Mihragān, important dates on the Muslim calendar could still hold precedence and aspects of the celebration of Mihragān needed to be adjusted or accommodated.

Returning to the comparison between the celebration of Mihragān and the 'Markets of the Arabs', the conflict between solar pre-Islamic calendars and the lunar Islamic calendar was a place of competition between pre-Islamic and Islamic cultural traditions. In the case of the 'Markets of the Arabs,' the seasonal movement of merchants between markets and fairs became disrupted by the Muslim pilgrimage which followed the lunar calendar.[50] In the case of Mihragān, we see conflicts as described above when the seasonal pre-Islamic holiday overlaps with an Islamic holiday determined by the lunar calendar. The move towards a lunar calendar further disrupted the rhythms of a solar calendar, as Bīrūnī lamented above.

Mihragān as Middle Ground

How did the celebration of Mihragān and particularly the practice of gift exchange act as a 'middle ground' between the Arab-Muslims and their Iranian subjects? We should return to Balkh for an illustrative case. Under the year 120/737, Ṭabarī relates an account—again on the authority of Madā'inī—related to the death of the Umayyad governor of Khurāsān Asad b. ʿAbdullāh al-Qasrī (d. 120/738) focused on a gathering in the city of Balkh for the celebration of Mihragān. During this gathering, the *amīrs* and *dihqāns*—or the military commanders and the petty landed gentry implying a gathering of both the imperial, largely Arab elites of the Umayyad caliphate and the local, Iranian hereditary elites—presented Asad with a variety of gifts so numerous they filled the cloth (*simāṭ*) that had been laid out for the banquet. Among these gifts were those specified as brought by the governor of Herat named Ibrāhīm b. ʿAbd al-Raḥmān al-Ḥanafī and the *dihqān* of Herat (whose name may itself be Khurāsān) which included two pavilions made of gold and silver (*qaṣr min fiḍa wa qaṣr min thahab*),[51] pitchers (*abārīq*) and platters (*ṣiḥāf*) of gold and silver, silk brocade (*dībāj*) from Marw, Qūhistān, and Herat, and a large gold ball.[52] Once these gifts had been presented, the *dihqān* of Herat rose and gave a speech before the gathered nobles in praise of Asad comparing him favorably to the pre-Islamic rulers

Amīn, 1970), 594.

48 Ibid., 273–4.

49 Ibid., 555–6.

50 Michael Bonner, '"Time Has Come Full Circle': Markets, Fairs and the Calendar in Arabia Before Islam', in: Asad Q. Ahmed, Behnam Sadeghi and Michael Bonner, eds., *The Islamic Scholarly Tradition: Studies in History, Law, and Thought in Honor of Professor Michael Allan Cook*, (Leiden: Brill, 2011), 13–48.

51 In his translation of Ṭabarī, Khalid Yahya Blankinship suggests these are models of fortresses, but that seems unusual. In his dictionary, Lane suggests that *qaṣr* may mean a pavilion or any kind of small enclosure as well as a fortress or fortified estate which better fits the context. Ṭabarī, *The History of al-Ṭabarī* 25 (trans. Khalid Yahya Blankinship; Albany, NY: State University of New York Press, 1989), 168, n. 586.

52 Ṭabarī, *Ta'rīkh al-rusul wa'l-mulūk*, vol. 2, pp. 1635–6; Ṭabarī, *The History of al-Ṭabarī* 25, 167–8.

of Iran (*maʿshar al-ʿajam*), noting Asad's leadership, his management of his household, entourage, and clients, his construction projects, and his recent military victories against the Türgesh Turks and the Arab rebel Ḥārith b. Surayj (d. 128/746).[53] Asad then proceeded to distribute his own honors. He gave the gold pavilion to ʿUdhāfir b. Yazīd (or Zayd) of the Banū Taym Allāh b. Thaʿlaba of the Bakr tribe of the Rabīʿa confederacy—the large tribal confederation of eastern Arabs who competed in the inter-Arab conflicts of the Umayyad period.[54] The silver pavilion was given to Maghrā' b. Aḥmar, described by Ṭabarī as both the head of Qays—the large tribal confederation of northern Arabs also involved in the inter-tribal conflicts—and of Qinnasrīn—one of the five military districts or *jund* of Umayyad Syria.[55] Other gifts were distributed to unnamed attendees (*fulān*) including an otherwise unknown Ibn al-Ṣaydā' who, when given the choice between two platters, weighed them in his hand to determine the heavier before Asad told him to take both. Others did not wait to be called on by Asad. Abū al-Yaʿfūr 'who used to go before the ruler of Khurāsān in military campaigns' and Maymūn al-ʿAdhdhāb or Maymūn the Torturer called attention to their service and they were given garments of silk brocade (*dībāj*) in return.[56] Not everyone was thrilled with Asad's behavior and the poet Nahār b. Tawsiʿah—who had served at the courts of many Umayyad governors of Khurāsān—attacked those gathered, saying 'You are few when one who rewards summons to war, but on the morning of al-Mihrijān you are many.'[57] At the end of this anecdote, Asad fell ill from an abscess (*dubaylah*) and died.

Almost 90 years after Asīd b. al-Mutashammis claimed ignorance of Mihragān, Asad led a complex and sophisticated celebration of the festival in Balkh. In Asad's performance, we can see how Mihragān developed as a site of 'middle ground' between the Umayyads and the local population. Here the celebration of Mihragān allowed for an exchange of gifts and the renewal of bonds not only between Asad and the people of Balkh but between the local population, Asad, and the Arab imperial and tribal elites more broadly. In order to understand this development, it is important to think about the historical context. Asad governed a Khurāsān that was in substantial disarray. He had served as governor twice, both times during the reign of the Caliph Hishām (r. 105–125/724–743). The first instance came in the year 106/724–5 in the midst of a particularly severe outbreak of inter-Arab tribal fighting during which a large number of Arabs refused to campaign alongside the governor of Khurāsān into the Ferghana Valley, choosing instead to hold back at Balkh which was subsequently besieged by a force loyal to the governor. The fighting ended only when Asad, himself of the Yaman or southern Arab tribal confederacy, was announced as the new governor. But he himself was removed from office just a couple years later (109/727–8) due to his excessive favoritism towards the Yamanites. Asad returned again in 118/736 after an Arab rebel from the Banū Tamīm named Ḥārith b. Surayj led an uprising against the Umayyads, possibly rooted in the Murji'ite religious movement. This uprising originated in Ṭukhāristān, first took Balkh, and then threatened Marw with an army of 60,000 men including a mix of Arabs and the local *dihqāns*. Following Asad's arrival, Ḥārith joined forces with the Türgesh Turks who had taken Transoxiana out of the hands of the Umayyads and were making regular incursions into Ṭukhāristān.[58] In short, Asad was

53 Ṭabarī, *Ta'rīkh al-rusul wa'l-mulūk*, vol. 2, pp. 1636–7; Ṭabarī, *The History of al-Ṭabarī* 25, 168–9.

54 On ʿUdhāfir's tribal lineage see Ibn Durayd, who describes him as having a fat neck (*al-ghalīẓ al-ʿunuq*). Muḥammad b. al-Ḥasan b. Durayd, *al-Ishtiqāq* (ed. ʿAbd al-Salām Hārūn; Cairo: Mu'assasat al-Khānjī, 1958), 353.

55 Maghrā' would later play an important role in the inter-Arab conflicts of the Umayyad era when he denounced the last Umayyad governor of Khurāsān Naṣr b. Sayyār (d. 131/748) before the Caliph Hishām (r. 105–125/724–743). For Maghrā''s actions against Naṣr, Ṭabarī, *Ta'rīkh al-rusul wa'l-mulūk*, vol. 2, pp. 1719–25; Ṭabarī, *The History of al-Ṭabarī* 26 (trans. Carole Hillenbrand; Albany, NY: State University of New York Press, 1989), 59–65.

56 Ṭabarī, *Ta'rīkh al-rusul wa'l-mulūk*, vol. 2, pp. 1637–8; Ṭabarī, *The History of al-Ṭabarī* 25, 169.

57 Ṭabarī, *Ta'rīkh al-rusul wa'l-mulūk*, vol. 2, p. 1638; Ṭabarī, *The History of al-Ṭabarī* 25, 169.

58 For the various conflicts that erupted in Khurāsān during the reign of Hishām, Khalid Yahya Blankinship, *The End of*

governor of a province that was severely divided not just between the conquered Iranians and ruling Arabs but also among the Arabs themselves, and Balkh was consistently at the center of these conflicts. During the Mihragān festivities described above, Asad's victories over Ḥārith and the Türgesh were specifically praised by the *dihqāns*. It is therefore attractive to think of the Mihragān celebration in terms of renewing the bonds between these various parties—with the local notables bringing Asad gifts and praising him for his ideal leadership which they compared favorably to the Sasanian model—but what happened in Balkh was not the practice of gift exchange that we previously saw attributed to the Sasanians.

Central to the Sasanian model discussed above—as reported in a variety of fourth/tenth century Arabic texts—was the reciprocity of gift exchange. On Mihragān, gifts were brought to the king and he remunerated with his own gifts and honors thus reaffirming the bonds between himself and his courtiers, clients, and relatives. We may think of this reciprocity in the terms of sociologist Marcel Mauss's obligations to give, to receive, and to reciprocate which—in Mauss's analysis—built a total spiritual bond among members within a society.[59] In Asad's case though, the gifts were not truly reciprocal. Instead, Asad acted as a transit point through which the gifts of Mihragān presented by the *dihqāns* were redistributed to the Arab notables. This followed a different model of gift exchange, one with its own cultural logic that focused on the obligation of the Arab tribal *shaykh* to generously redistribute his wealth among his followers and potential followers in order to insure their loyalty. It has been argued that during the Umayyad period this system of gift exchange moved from the realm of the tribe to that of the military retinue and that it became the job of governors and military commanders to employ their generosity on the model of a tribal *shaykh* in order to obtain and maintain the loyalty of their troops.[60] It is therefore perhaps worth noting that the first recipient of gifts from Asad was ʿUdhāfir b. Yazīd—a representative of the Rabīʿa tribal confederacy—while the second recipient was Maghrā' b. Aḥmar—the head of the Qays confederacy—both of whom rivaled Asad's own Yamanites. Having been removed from office once for his favoritism towards Yaman, it is valuable to see Asad so specifically recognize Rabīʿa and Qays and honor them with the most ostentatious gifts, the gold and silver pavilions.

The movement of the gifts of Mihragān from the *dihqāns* to the representatives of the Arab tribal confederations represented a shifting of political and economic power. Amidst the tumults of early second/eighth century Khurāsān, the *dihqāns* expressed their commitment to Asad and the Umayyads through the presentation of gifts on Mihragān—likely in an effort to preserve their own status though we are lacking the Balkhīs' voice and perspective on these events—and Asad expressed his role as *shaykh* among the feuding Arab tribes by redistributing those gifts among the Arabs. This then represents a 'middle ground' between the two parties in which neither group maintains fully their previous traditions, rather a hybrid form of gift exchange has developed with the governor at the center which allowed him to maintain portions of both preexisting practices, operating in both directions, and situating himself in the proper position of authority in the minds of both sides. In the broader scope, we may even think of the larger transformation of the gifts of Nowrūz and Mihragān into a form of non-canonical

the Jihād State: The Reign of Hishām Ibn ʿAbd al-Malik and the Collapse of the Umayyads (Albany, NY: State University of New York Press, 1994); Haug, *The Eastern Frontier*, 129–51.

59 Marcel Mauss, *The Gift: Forms and Functions of Exchange in Archaic Societies* (trans. Ian Cunnison; London: Cohen & West Ltd, 1966).

60 Patricia Crone, *Slaves on Horses: The Evolution of the Islamic Polity* (Cambridge: Cambridge University Press, 1980), 31–3; Michael Bonner, *Aristocratic Violence and Holy War: Studies in the Jihad and the Arab-Byzantine Frontier* (New Haven, CT: American Oriental Society, 1996), 4–5.

taxation in such terms. The Iranian notables brought the governors and/or caliph gifts which he then redistributed to his tribe/retinue. Instead of circulating within the single community though, the gift exchange now becomes a conduit by which wealth flowed from one party to another. Like a money laundering operation—in as much as it placed the gifts one step removed from their origins, not necessarily implying that this source was illicit—this transfer of gifts from one cultural milieu to another may also remove some of the problematic connotations found in the gifts themselves as the physical objects are transformed from symbols of Iranian kingship and Zoroastrian ceremony to signs of the Arab *shaykh's* generosity.[61] In the process, they ceased to be ceremonial objects imbued with cultural and religious meanings and were instead monetized.

Conclusions

By applying the idea of 'middle ground' to the practice of gift-giving during the celebration of Mihragān in the early Islamic period we can better understand this practice not as a simple two-way exchange and move beyond the notion of the gifts of Mihragān (and Nowrūz for that manner) primarily as a form of extra-legal taxation. Rather we can see the gifts of Mihragān as part of a complex intersection between Iranian tradition and Arab–Muslim rule. As such Mihragān was not a neutral site of celebration; instead it sat at an intersection between cultures, religions, and even calendar systems and it could become a site where they would compete with each other. While we are often lacking non-Muslim Iranian voices in these exchanges, relying instead on the accounts of (in most cases Iranian) Muslims writing centuries after the fall of the Sasanian Empire, and are therefore left primarily with the view from the rulers receiving the gifts, we can still see that both sides of the exchange altered their behaviors and expectations around the celebration of Mihragān to meet these new realities. Non-Muslim Iranians were active participants in this practice, not merely passive sources of revenues. Most telling of the way these traditions were being transformed, perhaps, is the manner by which governors like Asad b. ʿAbdullāh al-Qasrī could use Mihragān not just to build bonds with the local Iranian elites but also to reconfirm their relationship to the Arab military and tribal elites in a manner culturally recognizable to this separate constituency.

61 We may in this instance compare this exchange to the donation of 50 gold and silver idols taken from the temples of Kabul—likely some combination of Buddhist, Hindu, and pagan in origin—by the Ṣaffārid Amīr Yaʿqūb b. al-Layth (r. 247–65/861–79) to the Caliph al-Muʿtamid (r. 256–279/870–892) for the benefit of the welfare of the people of Mecca 'despite its association with the infidel.' By donating the idols to charity, specifically for the benefit of the Muslim community of Mecca, the taint of idolatry could be removed from the precious metals. Ṭabarī, *Ta'rīkh al-rusul wa'l-mulūk*, vol. 3, p. 1841; Ṭabarī, *The History of al-Ṭabarī* 36 (trans. David Waines; Albany, NY: State University of New York Press, 1991), 119; *Ta'rīkh-i Sīstān*, ed. Muḥammad Taqī Bahār (Tehran: Kitābkhānah-i Zavvār, 1935), 216; *The Tārikh-e Sīstan* (trans. Milton Gold; Rome: Istituto italiano per il Medio ed Estremo Oriente, 1976), 171–2.

Feasting in the Garden of God: Ramat Rahel and the Origins of Rosh Hashanah and Yom Kippur?[1]

JoAnn Scurlock

Introduction

Ramat-Rahel is a small settlement near Jerusalem. Archaeological excavations under the direction of Oded Lipschitz discovered there the remains of a Judean fort, which, after the campaign of Sennacherib to Judah, was rebuilt as a garden palace in which archaeological evidence indicates that a ritual was performed.[2] Apparently, there was a feast in the garden involving the king[3] and, to judge from the available space, his notables, at the end of which the remains of the meal were buried in a ritual pit discovered in excavation.[4] Above this garden palace was a second, somewhat larger, garden palace dating to the Persian period, into the floors of which a further set of ritual pits had been dug.[5] Remains of pollen in the plaster on the pool in the garden also gave precious evidence for what was planted there in the Persian period. We have olives and Lebanon cedar from the west, birch from the north, grapes, myrtle, poplar, willow, fig, and Persian walnut from the center and south, and lemons from the east.[6]

---

1 Abbreviations:
ARM VII = Jean Bottéro, *Textes économiques et administratifs* (Archives Royales de Mari 7; Paris: Imprimerie Nationale, 1957).
ARM IX = Maurice Birot, *Textes administratifs de la sale 5 du Palais* (Archives Royales de Mari 9: Imprimerie Nationale, 1960).
ARM XI = Madeleine Lurton Burke, *Textes administratifs de la sale 111 du Palais* (Archives Royales de Mari 11; Paris: Paul Geuthner, 1963).
ARM XII = Maurice Birot, *Textes administratifs de la sale 5 du Palais* (Archives Royales de Mari 12; Paris: Paul Geuthner, 1964).
ARM XXI = Jean-Marie Durand, *Textes administratifs des salles 134 et 160 du Palais de Mari* (Archives Royales de Mari 21; Paris: Paul Geuthner, 1983).
ARM XXIII = Guillaume Bardet, et alia, *Archives Administratives de Mari* 1 (Archives Royales de Mari 23; Paris: Éditions Recherche sur les Civilisations, 1984).
ARM XXIV = Philippe Talon, *Textes administratifs des salles Y et Z du Palais de Mari* (Archives Royales de Mari 24; Paris: Éditions Recherche sur les Civilisations, 1985).
ARM XXVI/2 = Dominique Charpin, et alia, *Archives Épistolaires de Mari I/2* (Archives Royales de mari 26/2; Paris: Éditions Recherche sur les Civilisations, 1988).
ARM XXX = Jean-Marie Durand, *Le nomenclature des habits et les textiles dans les textes de Mari* (Archives Royales de Mari 30; Paris: CNRS Éditions, 2009).
FM III = *Recueil d'études à la mémoire de Marie-Thérèse Barrelet* (eds. Dominique Charpin and Jean-Marie Durand; Florilegium marianum 3; Mémoires de N.A.B.U. 4; Paris: Société pour l'Étude du Proche-Orient Ancien, 1997).
FM XI = Grégory Chambon, *Les archives de Vin à Mari* (Florilegium marianum 11; Mémoires de N.A.B.U. 12; Paris: Société pour l'Étude du Proche-Orient Ancien, 2009).
FM XII = Antoine Jacquet, *Documents relatifs aux dépenses pour le culte* (Florilegium marianum 12; Mémoires de N.A.B.U. 13; Paris: Société pour l'Étude du Proche-Orient Ancien, 2011).
MARI 3 = *MARI Annales de Recherche Interdisciplinaires* 3 (Paris, Éditions Recherches sur les civilisations, 1984).
OECT 11 = Oliver R. Gurney, Literary and Miscellaneous Texts in the Ashmolean Museum (Oxford Editions of Cuneiform Texts 11; Oxford: Clarendon Press, 1989).

2 Diedre N. Fulton, et alia. 'Feasting in Paradis: Feast Remains from the Iron Age Palace of Ramat Rahel and Their Implications,' *Bulletin of the American Schools of Oriental Research* 374 (2015): 30-31.

3 A sherd showing a royal figure on a throne turned up at Ramat Rahel in Aharoni's excavations (Fulton et al.; 'Feasting in Paradis,' 40-41).

4 Fulton et alia, 'Feasting in Paradis,' 31-33.

5 Diedre Fulton, 'Feasting in Iron Age Judah: A View from Ramat Rahel,' Paper given at the Annual Meeting of the Society of Biblical Literature (San Antonio, 2016), Session S21-142.

6 Dafna Langgut, 'The Migration of Plants Within the Persian Empire,' Paper given at the Society of Biblical Literature Meeting, Boston, 2017, Session S18-233.

This represents a botanical tour of the world centered on Persia. Analysis of the animal bones from the sacrificial pits reinforces this impression. There were pigs and dogs (representing the cultivated land), camel (representing the desert), gazelle (representing the steppe) and ibex (representing the mountains). [7]

From what we know of ancient Persia, this structure may plausibly be identified as a Paradise garden. According to Bruce Lincoln, these gardens were places of feasting for the king and his nobles and part of a projection of Persian soft power. In addition, the presence of such gardens reinforced a cruxial nexus between the imperial god, The Wise Lord Ahuramazda, on the one hand and his chosen Shahanshah, King of Kings, and the empire on the other. They were known as Paradise gardens because they represented a gathering together of plants and animals, as the empire was of peoples, from all quarters of the world, the object being for The Wise Lord to use this collection to restore his original Creation consisting of one plant, one animal and one male Persian.[8] This desire for gathering the scattered is directly referenced in the plants planted in the Ramat Rahel garden and in the animals put into the sacrificial pits, which collectively represent flora and fauna of all regions and climatic zones known to Persia.[9]

It would be hard to argue, in view of the striking parallels of feasting activity involving the king and ritual pits being dug in a garden palace, that there is no connection between the Persian period ritual for which there is textual evidence and the Judean forerunner for which there is none. There are, however, clear differences as well--most strikingly the contents of the Judean ritual pits that do not have the same variety of mammals, that have an exceedingly unusual focus on immature animals, [10] and that have a wide variety of birds and fish,[11] all in contrast to earlier levels at the site[12] as well as to the Persian period equivalent pits.[13]

We do not know what was in the Judean garden, but it is exceeding improbable that there was any reference to the four quarters, let alone the entire Mazdean creation narrative. It is not, however, impossible that there is some connection, however, tenuous, between the Persian ritual and its Judaean counterpart. Judah, too, had a divinity considered to be the original Creator and was not the king his anointed? Judah had once been part of a multi-national empire and Josiah apparently had dreams of a restoration of rule from the Nile to the Euphrates. There was thus room in Judah for soft power reaffirmation of the relationship between king and notables that could profit from divine imprimatur in the form of celebration of original creation, when the earth brought forth the birds and the seas the fish and Adam and Eve lived in the Garden of Eden.

The immature animals[14] indicate autumn as the time chosen for this celebration, which raises a very interesting question. Could this be the origin of Rosh Hashanah, the seventh month New Years' Festival that to this day celebrates original Creation? This seems impossible at first

7 Fulton, 'Feasting in Iron Age Judah'.
8 See Bruce Lincoln, *'Happiness for Mankind': Achaemenian Religion and the Imperial Project* (Acta Iranica 53; Leuven: Peeters, 2012), especially pp. 3-104.
9 See above.
10 Fulton et alia, 'Feasting in Paradis', 34-35.
11 Ibid., 36-37, 42.
12 Ibid., 37, 39.
13 See above.
14 Fulton et alia, 'Feasting in Paradis,' 34-35. At three to six months, we have spring and summer lambs and we are in the fall.

blush. What does this festival have to do with human kings as opposed to the coronation of God as King and, for some, the celebration of a sacred marriage? [15]

Admittedly fish is on the menu for Rosh Hashanah, as apparently also for the feasting at Ramat Rahel to judge from the large number of fish bones recovered from the site,[16] but there is so much else to account for. What was the original motivation for the blowing of ram's horns at Rosh Hashanah? Why a whole fish, leeks and other vegetables, plus dates and pomegranates, to mention only the most obviously Near Eastern of the foods? Also demanding explanation is the symbolic casting of sins into a river or canal.

And then there is the mysterious connection between Rosh Hashanah, Yom Kippur and the Day of Judgment that ultimately allows for one more year of life inextricably mixed up with an even more mysterious Azazel ('goat of God') to whom a scapegoat was to be released in the steppe as an offering, according to Lev. 16.20-22, 26. Apart from the name, we know only that Azazel was one of the watchers, that is, a constellation, and that there is some connection to giants imprisoned at Babylon, either right side up or more usually upside down.[17] With Yom Kippur, a whole new set of problems opens up--how old is this festival and what, if any is/was its relationship to the ritual performed at Ramat Rahel?

We need more evidence. Fortunately, it is, in fact, ready to hand. The garden palace at Ramat Rahel dates to after Sennacherib's invasion, there being a completely different structure there in the preceding period.[18] This suggests the possibility of Assyrian influence here. It so happens that Sennacherib renovated a Middle Assyrian structure outside of Aššur and in the same relationship to that city as Ramat Rahel is to Jerusalem. This structure has been excavated, and it turns out to be a garden palace meaning that it did not simply have an attached garden, as was standard for Assyrian palaces, but that it actually centered around an enclosed garden. The excavator Andrae studied this garden and discovered that it consisted largely of pomegranate trees dug into the bedrock.[19] Pomegranates bloom twice a year, once in Spring and once in Autumn, and a long known Middle Assyrian Ritual of Kingship provides the obvious text for the performances of both seasons.[20] The Autumn festival involved feasting between the king and his nobles and the burial of food offerings in sacrificial pits. I think we have a match.

As luck would have it, the excavations at Mari turned up what appears to be the ultimate original of the Autumn ritual,[21] unfortunately fragmentary but with supplemental information

15 For details on this, and all other aspects of Rosh Hashanah, see https://en.wikipedia/Rosh_Hashanah.

16 See above.

17 John C. Reeves, 'Some Parascriptural Dimensions of the 'Tale of Hārūt wa-Mārūt,' *Journal of the American Oriental Society* 135 (2015): 827, 829, 831-832, especially 829 which adds: 'Therefore Israel brings offerings and casts one lot for the Lord, so that He might accept atonement for all the sins of Israel, and one lot for Azazel, so that he might bear the burden of the sins of Israel.'

18 Fulton, 'Feasting in Paradis,' 30.

19 Walter Andrae, *Das wiedererstandene Assur*, 2nd edition (Munich: C.H. Beck, 1977), 219-224. For illustrations, see Laura Kataja and Robert Whiting, *Grants, Decrees and Gifts of the Neo-Assyrian Period* (State Archives of Assyria 12; Helsinki: Helsinki University Press, 1995), 105, 107 figs. 19-20.

20 See below.

21 With Daniel E. Fleming, 'Chroniques bibliographiques 1. Recent Work on Mari,' *Revue d'Assyriologie* 93 (1999): 159-160, following Jean-Marie Durand and Michaël Guichard, 'Les rituels de Mari,' in *Recueil d'études à la mémoire de Marie-Thérèse Barrelet* (eds. Dominique Charpin and Jean-Marie Durand; Florilegium marianum 3; Mémoires de N.A.B.U. 4; Paris: Société pour l'Étude du Proche-Orient Ancien, 1997), 19-23, this ritual is not 'literary'. The reason the word is in quotes is that when it is used to describe a work of actual literature, say Shakespeare, it means something beautifully crafted and original. When used by cuneiformists, it usually means that the text is either a pack of lies or copied mindlessly with no thought of ever putting it to any practical use, or both at the same time. So historical texts are propaganda, not

from administrative texts and letters as well as one unexpected source. The Amorite king Šamši-Adad I had this (and others of his festivals)[22] celebrated on cylinder seals that were used by his officials to seal official documents, palace doorways, etc. Combined with later Assyrian (and, as we shall see, Babylonian) evidence from ritual texts and hemerologies, we are in a position to sketch the bare outlines of the history of a ritual. This will be a long journey, visiting cultic ceremonies performed in the same calendar month but in different times and places over the course of several millenia. All of these ceremonies have their own distinctive features; nonetheless there persists, as we shall see, a common core that sings a common tune.

The Invention of a Ritual of Kingship

There is archaeological evidence that Upper Mesopotamia experienced colonization from the overpopulated cities of Southern Mesopotamia in the pre-historic Uruk period.[23] This created a sort of Mesopotamian koiné, wherein parts of what is now Syria, including the ancient cities of Mari and Ebla, show clear signs of Sumerian cultural and even linguistic influence.

Šamši-Adad was an Amorite adventurer who, at the beginning of the 2nd millenium BCE, conquered for himself an extensive Upper Mesopotamian empire stretching across northern Mesopotamia from the Zagros mountains in the east to the Euphrates in the west, including not only Assyria, but bits of Babylonia and the Middle Euphrates region. With a view to cementing his rule, he instituted a ritual of kingship,[24] the full performance of which involved a community-wide celebration timed to coincide with a month that, from earliest Sumerian times, referenced original creation at Nippur, seat of the Sumerian Enlil of whom Šamši-Adad believed the god Aššur to be the Assyrian equivalent.

He was helped in this endeavor by the fact that this seventh month of the Nippur Calendar that was eventually adopted as the standard for all of Mesopotamia was also the month of the setting of astal Dumuzi (the constellation Aries), the first king and the favorite of the goddess Inanna/Ištar, in whose gift kingship lay. Last, but by no means least, the Autumnal Equinox was literally the turn of the year (ZAG.MUG) marked, in Sumerian times, by the heliacal rising of what we call Lyra but which was referred to as the 'Goat Star' and associated with the goddess Gula, mistress of the city of Isin, famous for its Sacred Marriage rites.[25] At Aleppo, under its

descriptions of historical events, laws written in stone more propaganda to show what a good king the king was and to cover the fact that he was nothing of the kind. In the case of ritual texts, 'literary' means that they were 'copied without intent to perform them'. We should not be having to fight this fight 100 odd years into this discipline.

22 With Fleming, 'Chroniques bibliographiques,' 161, there is no reason that all of the ritual texts found at Mari cannot be dated to the reign of Šamši-Adad.

23 A nice summary is given in 'Uruk Period,' *Wikipedia* (https://en.wikipedia.org/wiki/Uruk_period).

24 The seals of the Šakanakku period at Mari are typical presentation seals. They hint at some sort of relationship of the local king with Deritum in having an Inanna type figure presenting the king to her father Nanna with the symbols of Deritum's husband, presumably Ištaran, between them (See Oya Topçuoğlu, 'Emblems of Power: Ideology and Identity,' [PhD dissertation; University of Chicago, 2016], nos. 2-3 = Adelheid Otto, *Siegel und Siegelabrollungen* [Tell Biʾa—Tuttul 4; Wissenschaftliche Veroffentlichungen der Deutschen Orientgesellschaft 104; Saarbrucken: Saarbrucker Druckerei, 2004], 31-32, Tafel 40 a-i [M9]; 32-33, Tafel 41 a-d [M10]). When it comes to the Great Eštar, however, the king is shown being presented by an Inanna figure not to her father but directly to the month of her festival itself, represented by right side up and upside down figures (Topçuoğlu, 'Emblems of Power,' no. 5 = Otto, Siegel, 33-35, Tafel 42 [M12]). For more on this image, see below.

25 Pirjo Lapinkivi, *The Sumerian Sacred Marriage in the Light of Comparative Evidence* (State Archives of Assyria Studies 15; Helsinki: Helsinki University, 2004), 50-54, nos. 34-44.

Amorite rulers, the Autumnal Equinox was a time of determining destinies;[26] in Anatolia, Fall was the season for the celebration of the local divinity, a hunting god named Kurunta.[27]

In a polytheistic system, where cooperation between divinities is the norm, and outright equations of even the most implausible sort are also the norm, it was possible to create a kingship ritual in which everybody within the borders of a multi-ethnic empire could feel included, whether or not the gods whom they worshipped actually had their cult centers within the empire. Furthermore, some of Šamši-Adad's subjects were merchants residing in neighboring states. It will have been politic to enslure their safety by sending offerings to the neighbors' gods and by extending invitations to participate in a ritual in which they could be made to feel included.

Key to the system was the presence in all cultures of the Ancient Near East of goddesses of the Ištar type, spoiled brat daughters of the chief god of the pantheon with a decided tendency to play the field even when they were already married. For Mesopotamians, the shenanigans of Ištar had a great deal to do with the appearances and disappearances of the planet Venus, the goddess's astral manifestation. The festivals of Ištar in the cultic calendar were timed to coincide with key events in her life. This was not where these points fell in the calendar in any particular year, but the points at which these events will have occurred for the first time.[28] In this reconstructed first year ever, Venus appeared in the East in Ulūlu, remained visible in Tašrītu and Araḫšamna (months 6-8, i.e. the Fall) and then disappeared. She reappeared on the opposite horizon in Addaru and remained visible for Nisannu and Ayyaru (months 12-2, i.e. the Spring). In between, Ištar was in occultation.

The Festival of the Great Eštar at Mari: The Ritual Text

The key section of Šamši-Adad's festival of the 'Great Eštar' was recorded on a tablet of which a sizeable chunk is preserved. This begins on the night before the main body of the ritual, with a bed being set up in Eštar's house upon which the king may, if he wishes, sleep, a sort of optional pre-marital quickie. He had better make it quick, since he is roused before his usual wake-up time so that the goddess can get her breakfast. [29]

Everybody else in the temple has to get up, too, and they are ranged with Ninigizibara[30] facing Eštar, with lamentation priests to her left and a choir of male and female singers (the women behind the men) to her left and right. Also in attendance are brewers, carpenters, leather workers, carpet makers and weavers, expert craftsmen who lay out their tools, presumably to be blessed. Similarly, we have barbers with their razors. [31]

We are now ready for a curious rite involving a platter for *maṣḫatu*-flour, which is placed opposite Eštar. Not surprisingly, *maṣḫatu* flour is scattered on it, along with *sasqu*-flour moistened with

26 Mark E. Cohen, *Festivals and Calendars of the Ancient Near East* (Baltimore: CDL Press, 2015), 15.

27 For details, see Gregory McMahon, *The Hittite State Cult of the Tutelary Deities* (Assyriological Studies 25; Chicago: The Oriental Institute, 1991).

28 The movements of planets follow a cyclical pattern that repeats every so many years (different for each planet). If you wish to worship astral divinities that are not fixed stars, you either have festivals that dance round the calendar or you choose a particular cycle to which you peg your festivals, as later Astrologers would say, the planet's preferred points. We know from the actual cultic calendar that the latter choice was made.

29 Durand and Guichard, 'Les rituels de Mari,' 52, 57 no. 2 i 1'-6'.

30 For this goddess, see Durand and Guichard, 'Les rituels de Mari,' 47. She was apparently a by-form of Gula.

31 Durand and Guichard, 'Les rituels de Mari,' 52/54, 57 no. 2 i 7'-19'.

water. All around, bowls of water are set on the floor.[32] Meanwhile, in the countryside, as we know from the letters, bethels were being lain down (*ramû*) in the countryside.[33]

The next rite, for which the preparations are broken, is to be attended by Latarak[34] and a sort of ennead of other divinities, the goddesses being represented by their emblems. These sit or stand beside Eštar. The king dresses in a *lulumtum*-garment and sits behind the lamentation priests (so facing the goddess and her party). He is seated on a sailor's throne and attended by a single servant of the king's choice, who sits on a low chair. The remaining palace servants have to stand on either side and at a distance.[35] In this odd arrangement, may be recognized a wedding party, a Sacred Marriage in which the kingship of the king is annually renewed. There is also to be a wedding banquet ritually enacted by a designated eater.[36]

The lamentation Uru-am-ma-ur?!-ru-bi,[37] that is usually sung at the beginning[38] of the month,[39] is performed now, either in part or in its entirety depending on whether a *maḫḫu*-estatic (in tune with the divinities of Uruk) did (you do all of it) or did not (you do part of it) go into a trance at the beginning of the month. In either case, the choir has finished, and may leave.[40]

After a short break, we find the lamentation priests engaged in a cultic race. In the course of this race, they pass out the doors of the temple and then back in again, singing all the while. On their return, they sing a greeting song.[41]

Once this song is finished, the king stands respectfully for the Eršemma of the god Enlil (locally, Dagan), which is performed by a standing lamentation priest to the accompaniment of a *ḫalḫaltu*-drum. Grace having been said, we are ready for the meal. A designated eater sits by the head of the goddess and eats her portion, after which her plate is taken away by a designated away-taker.[42]

After the dishes are cleared, wrestlers engage in wrestling matches and *ḫuppu*-acrobats turn somersaults. Dirty (garments) are exchanged (for new ones).[43] We have more singing before the text breaks up. This must have taken a bit of time since, partway though the singing, the king gets to sit down.[44]

---

32 Ibid., 54, 57 no. 2 i 20'-28'.

33 For the last rite, see Durand and Guichard, 'Les rituels de Mari,' 32-35.

34 For this divinity, apparently a by-form of Nergal, see Ibid., 48.

35 Ibid., 54, 58 no. 2 ii 1'-18'.

36 See below.

37 For this lamentation, which is known from an Old Babylonian version, see Mark E. Cohen, *The Canonical Lamentations of Ancient Mesopotamia* (Potomac: Capital Decisions Limited, 1988), 536-603. The equation was recognized already by Miguel Civil, apud Durand and Guichard, 'Les rituels de Mari,' 49.

38 Durand and Guichard, 'Les rituels de Mari,' 58 unaccountably translate: 'end'.

39 At Mari, this will have been precisely on the 2nd day, to judge from the ritual schedule of the temple of Anu of Uruk where this lament originates. See François Thureau-Dangin, *Tablettes d'Uruk à l'usage des prêtres du temple d'Anu au temps des Séleucids* (Textes cunéiformes de Louvre 6; Paris: Paul Geuthner, 1922) 48: obv. 2-4.

40 Durand and Guichard, 'Les rituels de Mari,' 54, 58 no. 2 ii 19'-27'.

41 Ibid., 55, 58 no. 2 iii 6-13.

42 Ibid., 55, 58 no. 2 iii 14-21. For this interpretation, see Durand and Guichard, 'Les rituels de Mari,' 50-51.

43 For the interpretations, see Durand and Guichard, 'Les rituels de Mari,' 51-52. At Sippar, gods and goddesses changed clothing to coincide with the months of Spring and Fall, so 15 Addaru, 7 Nisannu, 10 Ayyaru, 3 Ulūlu, 7 Tašrītu, 15 Araḫšamna. In Uruk there were also changes for Summer (17 Duʾuzu and 1 Abu) and Winter (26 Kislīmu and 28 Tebeṭu). See Eiko Matsushima, 'On the *Lubuštu*-Ceremony of Bēl in the Seventh Century B.C.,' *Acta Sumerologica* 20 (1998): 111-112; Marc J.H. Linssen *The Cults of Uruk and Babylon: The Temple Ritual as Evidence for Hellenistic Cult Practices* (Cuneiform Monographs 25; Leiden: Brill/Styx 2004), 52-53.

44 Durand and Guichard, 'Les rituels de Mari,' 55, 58 no, 2 iii 22-33.

Finally, a series of three libations using water brought from the temple of the goddess is performed to the accompaniment of singing to honor first the goddess Eštar, then her brother Šamaš. The assembled divine guests get three libations to the right and three to the left. The king is given hand-washing water. The priests perform a further set of libations to goddess and her assembled guests. The god Šamaš receives no libations in this set. The rest is lost, apart from a note that water is to be supplied for the *maḫḫu*-estatics.[45]

The Festival of the Great Eštar at Mari: Administrative Texts

Administrative texts from Mari allow us to place this curious ritual exactly in time and add further details. The festival of the Great Eštar at Mari took place in the ninth month of the local Mari calendar. [46] From these texts, we may see that the first day, also the first day of the month, involved a welcome for the goddess and her attendants in the garden of the palace where she was treated to a sacrifice and presented with the first of two tables of sweet things. She may also have gotten a bath.[47]

After a gap on days two and three enlivened only by a possible further sacrifice in the garden of the king[48] and a mysterious issue of women's shoes to two priestesses and the king's daughters,[49] Eštar and her divine retinue moved into lodgings in the shrine of Belet-ekallim where she stayed from the 4th to the 6th.[50] The women presumably put the various garments that they had received to good use, but the men seem to have spent most of their time being bathed? with oil and drinking themselves blind, although there does seem to have been a *pagraʾu* offering to Dagan, since an issue of wine 'when Dagan' and the distribution of no less than 260 quarters of beef to the cooks are mentioned on the 6th.[51] This would also be appropriate to the prominent presence of an Eršemma of Enlil in the ritual.[52] Nergal and his chariot arrived on the 5th, since he received cedar oil on that day, usually used for baths.[53]

The busiest day, the height of the festival when everything came together was the 7th when the god Nergal got a *gibbu* in his temple[54] and his Chariot was honored with sacrifices probably, as we shall see later, outside in the open in the temple courtyard.[55] Meanwhile, there was a second round of sacrifice for Eštar in the garden of the king and she got the second of her two

45 Ibid., 55/57, 58 no. 2 iv 1-33, TL ii 1'-3'.

46 Bertrand Lafont, 1999. 'Sacrifices et rituels a Mari et dans la Bible.' *Revue d'Assyriologie* 93 (1999): 68 and Durand and Guichard, 'Les rituels de Mari,' 27-31 seconded by Fleming, 'Chroniques bibliographiques,' 160 follow Jack Sasson, 'The Calendar and Festivals of Mari During the Reign of Zimri-Lim in *Studies in Honor of Tom B. Jones* (eds. Marvin A. Powell, and Ronald A. Sack; Alter Orient und Altes Testament 203; Neukirchen-Vluyn: Neukrichener Verlag, 1979), 132-133, 134, 136) in assuming that the Great Festival of Eštar on Mari 9 and the Festival of Deritum of Mari 11 are the same festival. The mess this completely untenable assumption makes in our understanding of the cultic calendar at Mari (and Assyria of the same time period) is hard to imagine.

47 See below, Appendix 2 under Day 1.

48 There were either two consecutive sacrifices, or accounting was done on Day 2.

49 See below, Appendix 2 under Day 2.

50 See below, Appendix 2 under Days 4-6.

51 See below, Appendix 2 under Day 4-6.

52 See above.

53 See below, Appendix 2 under Day 5.

54 This has something to do with burning and is usually translated as 'holocaust'. Some of the better documented occasions suggest, however, that a torch-lit ceremony is what was meant. On the 18 Mari 11, there was a tour of the city on the part of acrobats, craftsmen and singers (FM III 74) followed by a torch-lit performance by the same team in the *sakannu* of the palace (FM III 95). An issue for what is probably the same event is for cedar oil for a *gibbu* from above the city to below the city (Dominique Charpin, 'Nouveaux documents du bureau de l'huile à l'époque assyrienne,' MARI Annales de Recherche Interdisciplinaires 3 [Paris, Editions Recherches sur les civilisations, 1984], no. 39).

55 See below, Appendix 2 under Day 7: Nergal.

tables of sweet things, this time with pistachios.[56] It is at this point that the wedding banquet will have taken place.

On the 7th, a sizeable number of sheep were contributed by the governors and tribal chieftains of the empire, giving the impression that communal feasting was contemplated. [57] This impression is confirmed by two undated issues that may be placed in this festival on several grounds. They are undated, which probably means that the described distributions were a once-a-year occurrence. Craftsmen, as in the brewers, carpenters, etc. who lay out their tools before Eštar in the ritual,[58] are among the recipients of ox shoulders, and there is a mention of the garden of the king as the point of origin of 32 sheep shoulders and who knows how many ox shoulders.[59] Sheep and cattle do not grow on trees, even at Mari. This meat must have been produced by a sacrifice in the king's garden, an uncommon venue for sacrifice in the Mari texts. These undated issues indicate that both fish and birds were also on the menu.[60]

At Mari, large distributions of expensive wine were usually sooner or later accompanied by what were referred to as *pudu*-offerings. Offerings of the same name appear in Old Babylonian Sippar where offerings of birds and fish, otherwise unusual concomitants of sacrifice, also appear. In AbB 7 159[61] (a letter with the heading Šamaš and Marduk, so presumably from Sippar), an official is trying to set up the yearly *pudum* offering for the brazier festival of Marduk and needs the recipient to buy and send him pronto 600 garlic cloves, 600 onions, 300 fresh water fish (*šarbūṭ*),[62] and 300 turtledoves (TU.GUR$_4$) of the two shekel size. Another letter, AbB 1 108,[63] also mentions a private *pudum* offering which the author is having to do for somebody else and he does not have what he needs: garlic, onions, fresh water fish (*šarbūṭ*) and turtledoves (TU. GUR$_4$).[64]

And, after all this, we are still not finished yet. Nergal gets yet more oil on the 9th, [65] and a flurry of offerings on the 10th that revolve round him, beginning with the *nubattum*, a reference, as we shall see, to food offerings deliberately left out overnight to spoil and cleared away the next morning,[66] so put out on the evening of the 9th and removed on the 10th. This was followed by another sacrifice in the king's garden and by the offering by barley soup of the type offered to ghosts. The gods entered the palace and Ea, Deritum and Annunitum got offerings, the latter twice, framing a special offering for the Chariot of Nergal.[67]

56 See below, Appendix 2 under Day 7: Eštar.
57 See below, Appendix 2 under Day 7: The Feast.
58 See above.
59 See below, Appendix 2 under Day 7: The Feast.
60 See below, Appendix 2 under Day 7: The Feast.
61 Fritz Rudolf Kraus, *Briefe aus dem British Museum: (CT 52)* (Altbabylonische Briefe in Umschrift und Übersetzung 7; Leiden: E. J. Brill, 1977).
62 For references, see *The Assyrian Dictionary of the Oriental Institute of the University of Chicago* S (Chicago: The Oriental Institute, 1984), 311 s.v . *sirbittu*.
63 Fritz Rudolf Kraus, *Briefe aus dem British Museum: (CT 43 und 44)* (Altbabylonische Briefe in Umschrift und Übersetzung 1; Leiden: E. J. Brill, 1964).
64 For more details, see Antoine Jacquet, *Documents relatifs aux dépenses pour le culte* (Florilegium marianum 12; Mémoires de N.A.B.U. 13 (Paris: Société pour l'Étude du Proche-Orient Ancien), p. 59; ARM XXIV p. 214.
65 See below, Appendix 2 under Day 9.
66 The dictionaries recognize that *nubattu* is from *bâtu* which means 'to spend the night' or 'to keep overnight' (see *The Assyrian Dictionary of the Oriental Institute of the University of Chicago* B (Chicago: The Oriental Institute, 1965), 169-173), but miss the references in late ritual texts where the offerings are left out overnight as in TU 39 rev. 9 (Linssen, *The Cults of Uruk and Babylon* 187, 190) where, contrary to custom, the meal is not cleared away but *ba-a-a-at i-ba-at*.
67 See below, Appendix 2, Day 10.

We are now beginning to understand this festival as one which was intended to spill beyond the confines of palace and temple at Mari. Contributions of garments by the districts of Mari, Terqa and Saggarum[68] and of sheep by the governors and tribal chieftains of the empire[69] made this a festival on an imperial scale. The curious rite with the putting down of bowls of water will have been visible only to a few, by anybody could see the bethels put down (*ramû*) in the countryside.[70] Similarly, the ritual meal with its eater as enactor and the washing of the king's hands afterwards will have been private, but a sizeable number of people will have been needed to consume the mountains of mutton, beef, sheep and birds laid out in the communal feast.[71]

Where are we in the Calendar for this Festival?

To get any further with our interpretation, it is essential to know when this festival will have been performed. See Appendix 1 for justification in terms of the agricultural cycle of the placement of the 9th month of the Mari calendar in the Mesopotamian 7th month of Tašrītu. In support of this identification, we know that the celebration in Šubat-Enlil, Šamši-Adad's capital, of the festival of Mari 8-10 was called an *elûnum*.[72] Ostensibly, this is related to *elû*, 'to rise up'[73] and *elēnu*: 'above' and refers to an appearance in the sky of a star or planet. Given this context, the mysterious tray of the Mari ritual[74] will have represented the Milky Way, and the bowls and bethels that were laid down, the stars of the night sky.[75]

As we have already mentioned, Venus's visibility was celebrated in Babylonian months Ulūlu, Tašrītu and Araḫšamna (months 6-8, i.e. the Fall) and again in Addaru, Nisannu and Ayyaru (months 12-2, i.e. the Spring). Understanding Mari month 1 as equivalent to Mesopotamian month 1, Nisannu, as is the case currently, would place the Festival of the Great Eštar in months 8-10 of the Nippur calendar or Araḫšamna, Kislīmu and Ṭebetu, simply impossible. Lining up Mari month 1 with Mesopotamian month 11, Šabaṭu, on the other hand yields an equivalency of Mari 8-10 with months 6-8 of the Nippur calendar or Ulūlu, Tašrītu and Araḫšamna, a perfect match for the Eastern goddess of the Autumnal Equinox.. In confirmation of this placement of the festival in Tašrītu, there are further sacrifices in the garden of the king on 28-29 Mari 9,[76] that would line up with the 28th-29th of Tašrītu in the Mesopotamian calendar. In Babylonia on the 29th of the month, as we shall see, lamentations with offerings were performed in open air settings to honor primordial gods.

In this context, the mysterious race and the coming in and out of doors of the Mari Eštar ritual[77] find a ready explanation. It is common around the world that rites of this sort--also tugs of war and wrestling matches--gravitate to celebrations of festivals that fall at the equinoxes, in this case the Autumnal one. We may also recognize in the acrobats turning somersaults,[78] the

68 See below, Appendix 2, Day 1.

69 See below, Appendix 2, Day 7: The Feast.

70 See above.

71 See below, Appendix 2, Day 7: The Feast.

72 Cohen, *Festivals and Calendars*, 281.

73 To note is that the term is often written very peculiarly, beginning with an 'è' (For references, see Cohen, *Festivals and Calendars*, 279-280). This is Sumerian for 'to come out', creating a pun with the Akkadian term from *elû*: 'to rise up'..

74 See above.

75 With Fleming, 'Chroniques bibliographiques', 161-162 and against Durand and Guichard, 'Les rituels de Mari,' 32-35, I do not see the bethels being laid down at Mari as having any funerary connotations.

76 See ARM XXI 18 for the 28th. Note also the *pudum*-offering on the 29th (M 10636 [Jacquet. *dépenses pour le culte*, p. 217]).

77 Durand and Guichard, 'Les rituels de Mari,' 55, 58 no. 2 iii 6-13.

78 Ibid., 55, 58 no. 2 iii 25.

'turn of the year'. Finally, the exchange of dirty garments for new ones is the sort of thing to be expected in ancient Mesopotamia at this time of year.[79]

What we have learned so far is that, at Old Babylonian period Mari around the Autumnal equinox, there was a festival which was accompanied by a communal banquet in which, among more usual fare, two unusual foods, fish and birds were served, accompanied by considerable imbibing of wine. The focal point of this rite was apparently a sacred marriage between the king and an Ištar-goddess, who was to grant him and/or to annually renew his kingship.[80] What is yet a mystery is the prominent role played by the lord of the Netherworld, Nergal, who would seem to intrude upon the proceedings like an unwelcome guest.

Insights from Comparative Material from Uruk

We may well wonder what the lamentation performed for Ištar of Uruk in Southern Iraq (in the temple of Anu at the beginning of the month) is doing being recited at Mari, of all places, in faraway Syria (in the course of the ritual of the Great Eštar).[81] To answer this question, we need to have a look at Uruk. We have a number of rituals from this city which are preserved in very late copies but which may still be presumed to contain some fossilized elements from much earlier periods. As luck would have it, the chief god of Uruk, the god of the heavens, Anu, had his main *akītu*-festival in Tašrītu, the Mesopotamian month that will have corresponded to the 9th month at Mari. Even more fortunate is that this latter ritual contained a sacred marriage for Nanaya, one of the by-forms of Ištar of Uruk.

In this Uruk ritual, from 1 Tašrītu to 8 Tašrītu, Anu's silver and gold chariots bring breakfast to Anu and his wife Antu and their court in the *akītu*-house.[82] This was the temple outside a Mesopotamian city where a god went every year with a view to celebrating his first ever establishment of a relationship with his human consituency. During this time, the sacred marriage of Nanaya is performed in the Eḫilikuga chapel, which is the bed-chamber of the Eḫiliana.[83]

The first part of the 7th of Tašrītu is spent by Anu, his wife Antu and their court in the company of lament-singers, regular singers and cooks who supply the bread, meat, milk and various alcoholic beverages served in a variety of vessels including *namḫaru*-vats; some of these were drinking vessels in the shape of ravens in honor of Corvus. This will have been for the wedding banquet of Nanaya.[84] The king is conspicuous by his absence in the rites of 7 Tašrītu; he shows up only for the rites of 8-9 Tašrītu. Armed with our Mari evidence, we know exactly what he will have been up to. The 7th day of the 9th month at Mari, the equivalent of 7 Tašrītu at Uruk is the day for the king's sacred marriage with Eštar; what this means is that the king will have been the imagined groom for Nanaya's sacred marriage, whose wedding banquet is being shared at a distance by her parents.

79 At Sippar, gods and goddesses changed clothing to coincide with the months of Spring and Fall. For details, see above.

80 The majority of the administrative texts that describe this festival are not from the reign of its creator, Šamši-Adad, but from his arch-enemy Zimri-lim, who established his rule in Mari after Šamši-Adad's death.

81 Durand and Guichard, 'Les rituels de Mari,' 54, 58 no. 2 ii 19'-20'.

82 Linssen, *Cults of Uruk and Babylon*, 184, 188 TU 39 obv. 1-3.

83 Ibid., 184, 188 TU 39 obv. 4-5.

84 Ibid., 185, 188 TU 39 obv. 7-10

That done, it was time to go home. Anu, Antu and Ištar are dressed in their new clothing. A bull is brought in between the curtains of Anu's private quarters (for Anu to stand upon), and there is a procession by road and boat with the *akītu*-house either locked (in Tašrītu) or opened (in Nisannu), plus we do not forget the morning and evening meal. The procession is performed as on the 7th of Nisannu.[85]

On the 8th of Tašrītu, the door of Anu's temple in Uruk is opened before Anu and Antu. Their vizier Papsukkal comes down to the courtyard to attend them. Anu is given his golden pole and holds a review of his weapons, suns and chariots. A host of gods including Adad, Sîn, Šamaš, Ninurta and even Lugalbanda and Ninsun join forces in the courtyard to form a small army. Conspicuous by his absense is Nergal, who would normally be with Adad at the head of the pack when any killing work was to be done, and who does get to tag along in Nisannu.[86] Anu has a meal of roasted meat, and he passes the plate round for his allies to take a share. They bring Anu out into the courtyard while one of those with access to the temple, girded with a belt, carries a royal scepter before him. Anu heads to the area between the curtains, and sits on his golden seat where he receives a sacrifice of an ox and a sheep whose heart and head are put before him. A plate of *maṣḫatu* flour (our friend from the Mari ritual) is dumped on the heart and a *maqû*-vessel of wine upended over the head.[87] We are looking at a battle, the winner of which may lay claim to dominion over heaven, what ancient Mesopotamians called the Anuship. Perhaps most obvious are the chariots and the army of gods who share a pre-battle meal. More significant is the offering of the heads. This offering echoes a ritual performed by Mesopotamian kings with the heads of defeated enemies.[88] This, and the royal scepter and golden seat indicate a victory for Anu.

Meanwhile in Mari, the 7th of the 9th month (the equivalent of Meopotamian 7 Tašrītu, found the Chariot of Nergal in the temple of Nergal, where a *gibbu* was performed, and Nergal himself got cypress and cedar oil, presumably for a pre-battle bath.[89] The 8th of the 9th month was greated with a thundering silence, and the 9th day has only an issue of oil for Nergal.[90] This would seem to suggest that things did not go so well for him.

Meanwhile in Uruk, 9 Tašrītu sees several repeats of his victory ceremony as Anu pays visits to the Dias of Destinies (where the fates of human beings are decided), then to the area between the curtains where he faces East. Adad and the other gods face him, and the goddesses face his wife Antu for the heart and head sacrifice; god and goddess wash their hands at the same time. The final destination for the day is the Dias of Kingship where Anu and Antu form a court from their supporters. A figure whose name is broken, doubtless the loser of the conflict, winds up imprisoned in Antu's courtyard. Also of note is that one of the sacrifical victims is cut open and the liver is put onto Anu's dias from whence it is picked up by the diviner and the priest of Adad, presumably for the performance of extispicy on the subject of the prisoners' fate. This seems to be favorable to mercy, since the singers sing about the house coming to be at peace. In the background we are gradually winding down to dinner which is, contrary to custom, left out

85 Ibid., 185, 188 TU 39 obv. 11-15.

86 Ibid., 201-202, 204 KAR 132 obv. i 12, [ii 2].

87 Ibid., 185, 188-189 TU 39 obv. 16-35.

88 See JoAnn Scurlock, 'The Techniques of the Sacrifice of Animals in Ancient Israel and Ancient Mesopotamia: New Insights Through Comparison, Part 2,' *Andrews University Seminary Studies* 44/2 (2006), 260.

89 For references, see below, Appendix 2 under Day 7: Nergal.

90 For references, see below, Appendix 2 under Days 8-9.

overnight.[91] In this 9th to 10th of Tašrītu overnight offering we may recognize the *nubattum* of exactly the same days in the equivalent 9th month of the Mari calendar.[92]

The 10th of Tašrītu at Uruk sees the clearing away of these offerings and is otherwise a day of rest. The final day, the 11th, sees one last shared meal between Anu and Antu and the gods who supported them.[93]

Given the timing, what we can say is going on in this Uruk ritual is that the god Anu in his astral manifestation as the plant Saturn, is having what later astrologers called his exaltation. To acheive this, he must come down from the Midst of Heaven Netherworld,[94] which will not happen without opposition. In short, Anu has to fight his way home to Uruk, and has mustered for this purpose divine allies who will form his future court.

The opponent who ends up imprisoned, with a possible release on the 10th, is most probably Nergal, lord of the Netherworld whose job it was to guard the Netherworld prison and to make sure that this truly was 'the land of no return.' We are also almost certainly dealing with the same Nergal who was having such a hard time of it in Mari, so there is a pretty direct connection between the Mari rites of the Chariot of Nergal that are attested in administrative rituals for 5-7 and 10 Mari 9 and those of Anu of Uruk in the equivalent month which accounts for the day by day coordination of cultic events in these widely separated centers.

One might, then suppose that while, in Uruk, Anu was preparing for battle on the 7th of Tašrītu, Nergal was doing the same in Mari, and that the rites of the 10th of the 9th month at Mari, which included both a *nubattum* and a sacrifice in the king's garden, as well as a group of supportive gods entering the palace[95] were designed to get Nergal out of the storehouse-prison in Antu's courtyard in Uruk and back home where he belonged before the dead broke open the gates of the Netherworld, and poured out to eat the living. What is more, the divination before Adad mentioned in this Uruk ritual[96] is likely to be the very Nabrûm that gave the 9th month at Mari its name in many of the Amorite calendars.

So, every year in the Fall, there was a celebration of the Amorite king's sacred marriage to the goddess Inanna/Ištar. This reached its apogee on the 7th of Mari 9 (Liliatum = Tašrītu) which featured Eštar's banquet against a backdrop of preparations for war between her father Anu and Nergal/Latarak. Before this, there was a sort of prequel when Eštar seems to have been made to feel very welcome but was otherwise left at loose ends[97] This interlude survives in the Hittite cultic calendar as a rite performed for three days of either spring or fall festival of Ištar of Nineveh. What is happening is that Ištar is being bathed by the queen and oiled and given things to eat and drink. This gives the queen the opportunity to 'drink the goddess' in private,

91 Linssen, *Cults of Uruk and Babylon*, 185-187, 189-190 TU 40 obv. 1-rev. 19, TU 39 rev. 1-9.

92 See above.

93 Ibid., 2004, 187, 190-191 TU 39 rev. 10-14, 15-25.

94 The Mesopotamian Netherworld was not actually nether at all, but lay in every direction, including up and down, from ego. For details on Mesopotamian cosmology, see JoAnn Scurlock, 'World Encircling River,' in *Irrigation in Early States: New Directions* (ed. Stephanie Rost; Oriental Institute Seminars 13; Chicago: Oriental Institute, forthcoming).

95 See below, Appendix 2, Day 10.

96 See above.

97 See above.

a time when a special personal favor may be asked.[98] And perhaps not only the queen; note the shoes issued to palace women on the 2nd of the 9th month at Mari. [99]

For what it is worth, later Assyrian hemerologies have the 3rd of Tašrītu as a 'no sex with the wife' day.[100] This ban also falls on the 6th when, at Mari, Eštar was in the shrine of Belet-ekallim.[101] This shrine, then, will have been the venue for the king's optional imaginary pre-nuptual intercourse with the goddess.[102]

More to be Learned from Mari Seals

More information may be gleaned from Šamši-Adad's celebratory seals. The scenes relevant to this ritual may be recognized by representations of the Goat Star,[103] curious figures representing the turn of the year,[104] and a cluster of grapes,[105] the grape harvest being in the Fall. From these seals, we learn that the *lulumtum* garment worn by the king for the occasion was essentially girded loins, or Amorite battle dress.[106] The king is not, however, shown grasping a weapon but instead an upside-down ram's horn, in other words, a shofar.[107] Similar horns are shown being waved about in the hands of celebrants[108] indicating that the object was to summon people to the wedding banquet. This battle dress and the wrestling matches[109] are, therefore, a battle of the sexes between the king and Eštar as well as a reflection of the conflict between Anu and Nergal that is going on at the same time. The presence of Latarak, a by-form of Nergal,[110] at the banquet of Eštar is probably not insignificant in this regard.

98 Maurice Vieyra, 'Ištar de Ninive.' *Revue d'Assyriologie* 51 (1957): 89-92, 95-96 no. 3.

99 See below, Appendix 2 under Day 2; with the possible exception of unedited texts, there are no entries for Day 3 of the festival.

100 For details, see below.

101 See above.

102 Durand and Guichard, 'Les rituels de Mari,' 52, 57 no. 2 i 2'-4'.

103 Topçuoğlu, 'Emblems of Power,' 97 fig. 4.2 = no. 19 = Otto, *Siegel*, 43-45, Tafel 50a-d and 51a-c (M23) shows this Goat Star framed by Pleiades on one side and Taurus being pushed up by a Laḫmu on the other. The effect is to entrap the Sumerian Autumnal Equinox in Araḫšamna between two representations of the Sumerian Vernal Equinox in Ayyaru.

104 Topçuoğlu, 'Emblems of Power,' seal no. 5 = Otto, *Siegel*, 33-35, Tafel 42a-f.

105 Topçuoğlu, 'Emblems of Power,' seal no. 158.

106 Compare Topçuoğlu, 'Emblems of Power,' seal no. 6 = Otto, *Siegel*, 53-56 Tafel 58a-d (M37) (ungirded) with Topçuoğlu, 'Emblems of Power,' seal no. 9 = Otto, *Siegel*, 36-38, Tafel 44a-g, 45a-g (M13) (girded). From *The Assyrian Dictionary of the Oriental Institute of the University of Chicago* L (Chicago: The Oriental Institute, 1973), 244a, we learn that the *lulumtum* was a 'battle cloak' and that these could weigh 4 2/3 minas. An administrator of the palace at Mari charged with preparations for the Feast of Eštar (ARM XXVI/2 285 r. 27') was having a seriously bad day. Not only did he have no *tiyāru*-cedar to make the chariot for the goddess to ride in and was having to use *ṣarbatu*-popular instead (ARM XXVI/2 285 r. 18'-31'), but he was short 2 minas of wool, almost half of what he needed for the *lulumtum*-garment and, to add insult to injury, the wool was an expensive import (ARM XXVI/2 285 obv. 3-18). From this text, we learn that the garment was woven from wool that was the color of *dūšu*-leather (ARM XXVI/2 285 obv. 5, 7, 8, 12, 13, 15, 16). This would be orange, essentially imitation leather armor.

107 Topçuoğlu, 'Emblems of Power,' seal no. 9 = Otto, Siegel, 36-38, Tafel 44a-g, 45a-g (M13). These are most clearly shown on Topçuoğlu no. 293 = Nimet Özgüç, *Acemhöyük—Burušhaddum I: Cylinder Seals and Bullae with Cylinder Seal Impressions* (Türk Tarih Kurumu Yayınları V/7; Ankara: Türk Tarih Kurumu Basımevi, 2015), 105 (CS 45).

108 Topçuoğlu, 'Emblems of Power,' seal nos. 294 & 344 = Özgüç, *Acemhöyük*, 101; (CS 40) and Özgüç, *Acemhöyük*, 123-24 (CS 65).

109 Durand and Guichard, 'Les rituels de Mari,' 54-55, 58 no. 2 ii 8', iii 23.

110 See Durand and Guichard, 'Les rituels de Mari,' 48. He appears in Durand and Guichard, 'Les rituels de Mari,' 54, 58 no. 2 ii 2'.

This banquet also appears on seals, with the king standing on one side of a table and the goddess on the other.[111] The food for this imagined banquet features a fish.[112] We know that in Ur in the Early Dynastic period, ocean fish were particularly to be eaten in the festivities of Ayyaru, the then Vernal Equinox.[113] Presumably, the then Sumerian Autumnal equinox in Araḫšamna will have featured fresh water fish to make a pair. Both types of fish are attested in the sacrificial pit at Ramat Rahel.[114]

The battle of the sexes involved the king getting what he wanted without having to put up for longer than necessary with Ištar's unwifely behavior. As with more standard Sacred Marriages, then, the wedding ended in a sometimes more and sometimes less amicable separation allowing both parties to marry other partners. This is symbolized on Šamši-Adad's official seals by the king's release of a bird representing Eštar that goes off to land on the hand of one of her imagined new partners, the Anatolian hunting god, Kurunta.[115]

Subsequently, the king will have renewed his vows with his human wife, thereby consummating his relationship with the goddess. In this context, the favor solicited by the queen during her three day wedding-preparation of Eštar[116] is likely to have been the all-important son and successor which it was her duty to provide to her husband. Both having a child and having a child of a particular sex were in Ištar/Inanna's gift.

As for the ex-wife goddess, she was, after her marriage to the king, temporarily married to other divinities--Adad of Aleppo for Amorite consumption and Kurunta for those of Anatolian origin. Both divinities are shown courting Eštar before her marriage alongside the king,[117] and both become her spouse, Kurunta after and Adad both before and after she had her temporary marriage with the king. Since both of these divinites already had wives, we may imagine this as somewhere between a fling and disastrous first marriage(s) that quickly ended in an amicable divorce. Either way, a relationship was created between the king and the involved divinities that long outlasted the sexually incontinent Ištar's love affairs, thus making the chief gods of two important ethnic constituencies within and without Šamši-Adad's empire royal patrons.

Feasting and Identity

Apart from divine patronage, the unstated goal of this kingship ritual was to create within the boundaries of Šamši-Adad's empire a sense of community and common identity that ignored ethnic boundaries, a nationalism of brotherhood as opposed to the otherhood of modern national identity. By its very nature, then, Šamši-Adad's kingship ritual had no intrinsic connection to this monarch or his dynasty and was eagerly adopted by those who eliminated his line, both at Mari and in Assyria. Amorite Zimri-Lim's short-lived attempt at continuation perished in the flames of his palace at Mari. Fortunately, the Assyrians, despite a campaign of extirpation directed at Šamši-Adad's descendants, embraced this ritual as foundational to their national identity and found other means to create a sense of brotherhood in their much

111 One of the seals, Topçuoğlu, 'Emblems of Power,' seal no. 324 = Özgüç, *Acemhöyük*, 161-62 (CS 97) even shows the designated eater described in the ritual as a tiny man standing before the king and mimicking his gestures.

112 Topçuoğlu, 'Emblems of Power,' seal no. 321 = Özgüç, *Acemhöyük*, 160 (CS 95).

113 Cohen, *Festivals and Calendars*, 74-75. For the position of the equinox in Sumerian times, see JoAnn Scurlock 'Nippur's Galileo Problem,' in this volume.

114 Fulton et alia, 'Feasting in Paradis,' 36-37.

115 Topçuoğlu, 'Emblems of Power,' seal no. 340 = Özgüç, *Acemhöyük*, 127-28 (CS 68).

116 See above.

117 Topçuoğlu, 'Emblems of Power,' seal nos. 328, cf. no. 325 = Özgüç, *Acemhöyük*, 122-23 (CS 64) and 149-51 (CS 85).

extended empire. These means included moving people from one side of the empire to the other and drawing young men in small numbers from all over to participate in joint military campaigns and joint building projects culminating in communal dinners in which herbs and spices from every corner of the empire were put together to create a communal curry. The message was everywhere the same. Keep your gods, keep your customs, but you need to think of people with different gods and different customs as your brothers engaged in a common enterprise from which all will benefit.[118]

The Middle Assyrian Rituals of Kingship

As for the kingship ritual, fast forward from Šamši-Adad I, contemporary of and kinsman of Hammurapi of Babylon at the beginning of the 2nd millenium to its end, the Middle Assyrian and contemporary Middle Babylonian period. As we have already noted, the Middle Assyrian Ritual of Kingship provides the obvious text for the performances of the rites associated with the twice yearly blooming of the pomegranate trees planted in the garden of the *akītu*-house in the steppe outside the city of Aššur rebuilt by Sennacherib to accommodate festival banquets that had, while the Middle Assyrian structure had lain in ruins, been celebrated in a garden inside the city.[119] The rituals in question were Sacred Marriage rites involving three separate marriages.

The first of these does not concern us. The second marriage performed annually in the Middle Assyrian period in Aššur's *akītu*-house in the steppe was between the Assyrian Enlil and Ištar by virtue of which she became Queen of Heaven. It was known as the bed ritual of the *kulīlu*, a silver diadem worn by goddesses on their wedding days. This rite will have been taken over from the Vernal Equinox marriage of Enlil with Annunītu[120] celebrated in Assyria by virtue of the equation of Aššur with Enlil by Šamši-Adad.

The Assyrian king, playing the role of paranymph, brought Ištar and Aššur into the bedroom, soaked the ritual bed with the contents of a *maqû*-bowl to represent consummation and offered a midnight snack of boiled meat to Aššur and roasted meat to Ištar. The bed also considerately received roasted meat as its reward. That done, the king came out where the assembled crowd could see him.[121] The rest of the ritual is unfortunately lost.

The final ritual was a marriage between the human king and the goddess which, in addition to all else, created an indirect relationship between the king and the national god. The beginning is unfortunately broken, but when we pick up the action, we have an *apu*-pit, which is receiving blood, honey and herbs, beer and wine. The prominence of this pit, used for making offerings to Netherworld and primordial gods, marks the timing of the ritual as being in the seventh

118 On feasting as a political tool used by the Assyrians 'to unify the conquered lands and to turn local elites into loyal clients', see also Fulton et alia, 'Feasting in Paradis,' 29; cf. 42-44 with further references. The emphasis on elites is something of an obsession with anthropologists generally as is the tendency to attribute the worst possible motives to empires generally and the Assyrian empire in particular (as in Dietler, apud Fulton et alia, 'Feasting in Paradis,' 2, 30). Those attending Aššurnaṣirpal II's banquet at Kalḫu were anything but elite and a good reign, by Assyrian standards, was one in which the notables spoke and common people listened **and** viceversa. Ernst Weidner, 'Assurnasirpal in Assur,' *Archiv für Orientforschung* 13 (1939/41) 210 no. 3 obv. 12-13.

119 Kataja and Whiting, *Grants, Decrees and Gifts*, no. 86.

120 This originally took place on the 10th of the month Ayyaru but was moved in Assyria to (the 10th of) the month Nisannu to accord with the progression of the equinoxes. Sennacherib states that the celebration at the beginning of the year that should take place in the *akitu*-house in the steppe outside Aššur was to take place in Nisannu (Kataja and Whiting, *Grants, Decrees and Gifts*, no. 86: 5, 14). For more, see Scurlock 'Nippur's Galileo Problem,' below in this volume.

121 Eric Ebeling, 'Kultische Texte aus Assur,' *Orientalia NS* 21 (1952) 143,145 ii 10-26.

month, Tašrītu. The king is sitting on a stool waiting for meat to roast. When it has, the king skewers the flesh of the front of the neck with a small iron dagger to feed the Lisikkutu gods while singers sing: Roast Meat, Roast Meat, Roast Meat, Let them eat'. This piece of meat goes into the *apu*-pit. There is a table, which gets purified before the king makes another public appearance.[122]

After a further lacuna, we find the king emptying a golden *maqû*-bowl belonging to the *ḫarû*-vat whose opening symbolized the defeat of Tiamat in the Assyrian version of the Babylonian New Year's Festival.[123] He stands on a pedestal while the singers sing: O, *illūru*-berry, O *illūru*-berry', a reference to rouged cheeks, to the accompaniment of cymbols held by the king and magnates. The king sits down at table for the *naptunu* (a communal meal in which king and notables feast together). After the meal, the *šangu*- priest enacts a mock battle (a dim reflection of the Nergal—Anu battle seen in Uruk and Mari but probably actually meant to reference a rather different battle celebrated, if that is the word for it, at Middle Babylonian Nippur).[124] The singer performs his rites and, when he has released the francolin (the bird whose release at Mari left the goddess free to move on to other partners). The *šangu* presents water for hand-washing to Ištar and the king. In Šamši-Adad's original, there was also a ritual hand washing of the king and his ex-bride that similarly took place after the meal; you might say that the king was washing his hands of her. In the Middle Assyrian ritual, the singer finishes his rites, after which the small audience of common people who have attended the ritual leave, and they shut the gates.[125]

The king takes a foreleg from the table (the word used is not Akkadian but *dura'u* = Hebrew *zerah*) and puts it on bread to feed the Lisikkutu while the singers sing: 'Who can spend the night in the Silver House?' This offering, too, ends up in the *apu*-pit with honey, herbs, beer and wine poured after it to wash it down. The singer fills in the *apu*-pit (with the rest of the remains of the meal) and the king puts his foot on the (filled in) *apu*-pit, stamping it down.[126] In the context of the seventh month, the remains in the pit represents a *kispu*-offering[127] to placate the primordial gods of original creation whose defeat in the time before time ensured the cycle of life and death in the time-bound world.[128] The rites completed, the king kisses [ground] and goes straight to the palace.[129]

The Middle Assyrian Tašrītu Sacred Marriage rite has room for the Mari pre-marital-sex-on-the-nearest-bed encounter in its broken section. However, as at Mari, the bed has no part to play in (what is preserved of) the rest of the ritual. Moreover, there is a ritual releasing of

122 Ibid., 143-144, 145-146 iii 1-28.
123 'The king who opens the *ḫarû*-vat in the race is Marduk, who defeated Tiamat with his penis' (Alasdair Livingstone, *Court Poetry and Literary Miscellanea* [State Archives of Assyria 3; Helsinki: Helsinki University Press, 1989], no. 37: 18).
124 See below.
125 Ebeling, 'Kultische Texte,' 144, 146 iv 1-14.
126 Note that the pits at Ramat Rahel were literally beneath the floor and 'not distinguishible above the courtyard level' (Fulton, et alia, 'Feasting in Paradis,' 31). The holes deliberately pierced at the base of some of the buried vessels is not a feature of feasting events (so Zuckerman, apud Fulton et alia, 'Feasting in Paradis,' 34) but a clear indication, attested round the world, of offerings intended for otherworldly recipients. Ditto the horse and rider figurines found with the vessels.
127 For these funerary offerings for the ghosts of humans and gods, see Akio Tsukimoto, *Untersuchungen zur Totenpflege* (kispum) *im alten Mesopotamien* (Alter Orient und Altes Testament 216; Neukirchen-Vluyn: Neukrichener Verlag, 1985), and for this set of such offerings, note especially pp. 201-217.
128 Is it completely irrelevant that Ramat Rahel is situated in the Rephaim Valley (Fulton et alia, 'Feasting in Paradis,' 30)?
129 Ebeling, 'Kultische Texte,' 144, 146 iv 15-29.

the francolin, which, in the original Šamši-Adad ritual, allowed Ištar to go where she pleased. Most interestingly, the answer to the singer's question: 'Who can spend the night in the Silver House?' seems to be 'Not the king', since afterwards he 'goes straight to the palace.' The obvious conclusion is that the king's relationship with the goddess was consummated with his queen after his return home at the end of the rite. This, as we have already suggested, will have been the ultimate finish to the original rite at Mari.

Battles in Middle Babylonian Nippur

The Mari rites of the seventh month crossed the battle of the sexes with a cosmic conflict centered around Anu of Uruk and the Chariot of Nergal. In Babylonia in the Middle Babylonian period, the seventh month was also associated with battles in the sky, but over a rather different issue that pitted the city of Nippur against Babylon.[130] We find ourselves in the seventh month in the middle of three confrontations between Ninurta (as the planet Mercury) and Marduk (as the planet Jupiter) over which of them would be Lord of Time.[131] As Nippur understood things, Ninurta won round one in the second month, Ayyaru, and Marduk round three in the ninth month, Kislīmu. The middle seventh month, Tašrītu, was serious bad news, since both parties lost and wound up in the Netherworld to be rescued by the goddess Gula, to whom offerings were correspondingly to be made.

These rites of Nippur have to be pieced together using an actual Middle Babylonian Offering List from Nippur paired with a Neo-Assyrian text known as the Nippur Compendium and a late Cultic Commentary. The Nippur Compendium's cryptic entries for the later activities for the seventh month are: Chariot of Enlil, Foremost Weapon of (Ninurta and) Enlil = For the rising of Šamaš (Renewal) and Urunakku = Reed hut[132]

The Middle Babylonian offering list may be lined up with these entries with the help of the preceding lines in the Nippur Compendium which refer to the Festival of Ištar involving a 'Long River Ordeal.'[133] This corresponds to the Offering List's final entry for the sixth month, 28 Ulūlu: 'She puts her garment down on the dias and (goes to) the River Ordeal.'[134]

The following entry in the Offering List is unfortunately broken, but it gives 7 Tašrītu as the date for the rite described in the Nippur Compendium in which the Chariot and Weapon were renewed by being left out overnight for the sun to rise over them.[135] We may perhaps assume that this is also what happened to the Chariot of Nergal at Mari. In any case, we are preparing for a battle. The opponent is revealed in the Offering List's next entry for 8 Tašrītu which records the issue to the Temple of Marduk of offerings appropriate to a procession to coordinate with the Nippur Compendium's Reed Hut rite.[136] The next day, 9 Tašrītu in the

130 For more on this see Scurlock, 'Nippur's Galileo Problem,' below in this volume.

131 This is fallout from Nippur's Galileo Problem for which see below. To note is that the 3rd millenium Lagash/Girsu calendar has an Ayyaru to Simanu festival and an Ulūlu to Tašrītu festival in which offerings are made to the planet Jupiter (Cohen, *Festivals and Calendars*, 62-63, 64-66).

132 Nippur Compendium § 13 29'-30' (Andrew R. George, *Babylonian Topographical Texts* [Orientalia Lovaniensia Analectia 40; Louvain: Peeters, 1992], 154-155).

133 Nippur Compendium § 13 26'-28' (George, *Topographical Texts*, 154-155).

134 Jonathan S. Tenney, 'The Elevation of Marduk Revisited: Festivals and Sacrifices at Nippur,' *Journal of Cuneiform Studies* 68 (2016), 165, 166 rev. 6. This is reading BÁRA TÚG.GAR.RA rather than Tenney's DÚR.GAR.RA. It is her garment that is going on the dias and not her posterior.

135 Tenney, 'Elevation of Marduk,' 165, 166 rev. 7 = Nippur Compendium § 13 29' (George, *Topographical Texts*, 154-155).

136 Tenney, 'Elevation of Marduk,' 165, 166 rev. 8 = Nippur Compendium § 13 30' (George, *Topographical Texts*, 154-155).

Offering List, has issues identical to those given to the Temple of Marduk at Marduk's *akītu* festival in the first month, Nisannu, but they go instead to the Temple of Gula.[137] The 26th of Tašrītu brings offerings described as the ŠUK LUGAL, the 'royal portion'.[138] The quantities and destination are comparable with Ninurta's release of captive gods on the 26th of the second month,[139] the triumphant finish to Ninurta's Ayyaru battle with enemies about which we will hear much more later.[140] The resulting reaffirmation of the kingship of Ninurta's father Enlil also reaffirmed the kingship of his human counterpart, which is why he receives a portion. The final rite of the month, on 29 Tašrītu, is an offering to the Primordial Mound, about which more presently.[141]

The Cultic Commentary, OECT 11.69+70, is unfortunately fragmentary but it adds in the crucial details. It, too, has the 28 Ulūlu entry about Ištar and her River Ordeal.[142] The following entry, the first for Tašrītu, is missing its date but this can, with confidence, be restored as the 7th. On this day, there was a lament for Lugaldukuga, 'king of the Primordial Mound' a forefather of Enlil, who was honored with a ritual *apu*-pit. This is the Nippur equivalent of the pits in the Middle Assyrian Kingship Ritual and of those whose archaeological remains we have been studying.[143] On the next day, which may be restored as the 8th, there was a purification rite for the people and Lugaldukuga was seated, presumably in the reed hut mentioned in the Nippur Compendium.[144] The 8th was also the day for the procession from the Marduk Temple, one would think taking Marduk to his *akītu*-house as in the Babylonian New Years' Festival in the first month of the year.[145] The procession was indeed followed on the 9th by offerings appropriate to an *akītu* of Marduk but they were made instead to the Temple of Gula which suggests that Marduk's journey was to Gula's house.[146] As to why this was necessary, the Cultic Commentary originally gave an explanation of which what survives is Gula and 'their ghosts',[147] quite enough to tell us that both Ninurta and Marduk wound up at the end of their battle in the Netherworld and were only rescued from this predicament by a goddess of healing whose Sumerian name, Nindinugga, means 'mistress who brings the dead to life.' The final entry in the Cultic Commentary for the month, 2[9] Tašrītu, has a lamentation, presumably to accompany the *kispu*-offerings to the Primordial Mound.[148]

In the Neo-Assyrian period, our information about the final version of the Babylonian celebration becomes more or less complete. From a Neo-Assyrian letter to the king, we learn that, in Babylon, Bel (Marduk) was dressed on the 7th of Tašrītu in a special garment. We also hear that the gate of the Esagila temple was left open on the 8th and that the divine statue went in procession 'as in the month Nisannu.'[149] What the king was not told, although he will certainly have known this, is that Marduk was being readied on the 7th for his ill-fated battle

137 40/40/2 for the Nisannu *akītu* of the Marduk Temple (Tenney, 'Elevation of Marduk,' 165, 166 obv. 12') and 40/40/2 for the Temple of Gula on 9 Tašrītu (Tenney, 'Elevation of Marduk,' 165, 166 rev. 9).

138 Tenney, 'Elevation of Marduk,' 165, 166 rev. 10.

139 Ibid., 165, 166 obv. 18'.

140 See below, Scurlock, 'Nippur's Galileo Problem,' in this volume.

141 Tenney, 'Elevation of Marduk,'165, 166 rev. 11. See below.

142 OECT 11.69+70 ii § 7'.

143 OECT 11.69+70 ii § 8'.

144 OECT 11 69+70 ii § 9'.

145 Tenney, 'Elevation of Marduk,' 165, 166 rev. 8.

146 Ibid., 165 rev. 9.

147 OECT 11 69+70 ii § 10'.

148 OECT 11 69+70 ii § 12'.

149 Simo Parpola, *Letters from Assyrian and Babylonian Scholars* (State Archives of Assyria 10; Helsinki: Helsinki University Press, 1993), no. 253 rev. 2-7.

with Ninurta and that the procession of the 8th was indeed as in the month Nisannu, only it was to gain the intercession of Gula for his release from the Netherworld prison. We may also recognize in the garment of the 7th of Tašrītu, the *lulumtum*, Amorite battle dress for an Amorite god.

The Late Babylonian (so Neo-Assyrian or Neo-Babylonian) ritual for the Esagila complex for the third encounter in the ninth month, Kislīmu, adds further details. On the [15th of Ayyaru], the 8th of Tašrītu and 4th of Kislīmu, Ninurta and the image of Bel got up from their cella's and took up a station before the [...] gate. [150] For Tašrītu, this will have been in the night preceding the eighth day in the open under the night sky to allow the sunrise to 'renew' the participants. The same will have been true of the 15th of Ayyaru and the 4th of Kislīmu.[151] The *lillisu*-drums that have been going on all night are relieved in the morning by *alû*-drums arranged as on 8 Tašrītu, when the gate before Bel is opened. With a brief interruption so that Marduk does not miss his morning meal, the rite of the *alû*-drums arranged as on 8 Tašrītu is performed. A golden *namḫaru*-vat is placed between them. This was destined to be filled from four *namzītu*-vats lurking in the left of the courtyard. These *namzītu*-vats are used to make beer by pounding beer mash and adding water, the distant ancestor of Budweiser.[152] This was, of course, to be allowed to ferment in the *namzītu*-vats. In the actual 8 Tašrītu equivalent, given the timing to coincide with the grape harvest, the participants could instead have used wine with grapes and brewers' yeast substituted for the beer mash. In any case the effect was similar to the bowls of water of the original Šamši-Adad star summoning ritual.[153]

For his son Šamaš-šum-ukin, Esarhaddon formalized royal participation by the king of Babylonia in the rites of the seventh month in Southern Mesopotamia. Although the destruction of Assyria was unspeakably violent, this Assyrian ritual continued to be performed by Neo-Babylonian kings. Our latest copy of the ritual instructions is from the reign of Artaxerxes.[154]

This festival, like its Assyrian counterpart celebrated in the seventh month in Northern Mesopotamia, was a pair with the rites of the first month of the year[155] as well as being part of what was originally a Sumerian festival honoring the Primordial Mound of Creation,[156] to which, as we have seen, offerings were still being made. Indeed, the end of the seventh month found late Babylonian householders making propitiatory offerings to primordial gods such as Lugaldukuga ('king of the primorial mound') and Enmešarra.[157] This Primordial Mound corresponds to the first stage in the evolutionary process described in the Babylonian Creation Epic in which there appear suddenly, by spontaneous generation, creatures neither living nor exactly dead but more or less both at the same time and located, if that is the word for it, somewhere and nowhere in infinite time and space. After the god Marduk got through battling with them, they became what we call the constellations. The constellations were forced by

150 G. Çağırgan and Wilfred G. Lambert, 1991-93. 'The Late Babylonian Kislīmu Ritual for Esagil,' *Journal of Cuneiform Studies* 43-45 (1991-93): 100: 164-166.
151 There was an offering on this date at Nippur (Tenney, 'Elevation of Marduk,' 165, 166 rev. 18) which is broken, but has the same amounts as on the 9th of Tašrītu (40/40/2) when Gula got her thank-you offerings.
152 Çağırgan and Lambert, 'Kislīmu Ritual,' 95-96: 55-62.
153 See above.
154 Claus Ambos, *Der König im Gefängnis und das Neujahrsfest im Herbst: Mechanismen der Legitimation des babylonischen Herrschers im 1. Jahurtausend v. Chr. und ihre Geschichte* (Dresden: Islet Verlag, 2013), 195.
155 For a chart of the events of what he styles The New Years' Festival of the Fall, see Ambos, *Der König im Gefängnis*, 54.
156 For details, see Blahoslav Hruška, 'Zum 'Heiligen Hügel' in der altmesopotamischen Religion,' *Wiener Zeitschrift für die Kunde des Morgenlandes* 86 (1996) 161-175.
157 Réné Labat, *Hémérologies et Ménologies d'Assur* (Paris: Librairie d'Amerique et d'Orient, 1939), 120-123: 18-21.

Marduk to move along tracks laid out for them, being killed and brought back to life in a yearly cycle that produced and maintained the flow of time.[158] At each repeated cycle, however, these primordial beings were given a chance, in the mutually destructive conflict between Ninurta and Marduk of the seventh month, to stop the clock, preventing the creative process that disturbed their sleep from ever having happened.[159] Since humans were part of that process, it was obviously in their interest to make sure that every time was like the first time.

On the 4th of Tašrītu, the king was to release seven marsh birds. The householder in the know also captured and released a bird on the 3rd or 4th, depending on which hemerology he consulted.[160] The accompanying recitation in Esarhaddon's ritual indicates that these were birds of heaven created by Anu whose release was meant to produce a similar release of the king at the hands of the Sungod, Šamaš.[161] This is a rather dramatic re-interpretation of the original bird release, but it quite suits its new context. We are here seeing a rather different side of the god of heaven, Anu. In the *Enuma eliš*, he is a creation of AN.ŠÁR,[162] representing the outermost of the spherical three heavens and earths that separated bounded space from infinity, and started the process of active Creation.[163] The date of Anu's birth, however, remains in the same month, Tašrītu, as Saturn's imagined exaltation.[164] Active creation as procreation produces successive generations of noisy, rambunctious gods. Anu's children and grandchildren are soon in need of champions, being in danger of being eliminated by the reactionary primordial gods whose greatest ambition is to spend eternity asleep. Unfortunately, in Tašrītu, the god Marduk was, depending on how you look at it, either not yet born or out of commission, and so the human king had to go to the Netherworld along with Marduk, hopefully both to be released with the intercession of the healing goddess Gula.

In preparation for this hazardous journey, the king underwent rituals of cleansing and purification seconded in both Assyria and Babylonia, by popular house cleanings. In the Cultic Commentary, these are mentioned only for 8 Tašrītu and associated with Lugaldukuga and the reed hut. Elsewhere, in both Assyria and Babylonia, house cleaning began already with the first day of the month and involved increasingly onerous taboos culminating in what is called in Hebrew the *shebat shebatin* (the seventh day of the seventh month) when just about any conceivable activity was forbidden.[165]

The calculation works well with seven days of Creation, but the Jewish Sabbath that this story justifies has its likely origins at the chthonic end of Yahweh's activites as Lord of She'ol. At Mari, the seventh month was at the Summer solstice, a time when funerary offerings were made to the ghosts of the dead. It seems obvious that food taboos of one kind or another would accrue

158 See JoAnn Scurlock, 'The World Encircling River,' in *Irrigation in Early States: New Directions* (ed. Stephanie Rost; Oriental Institute Seminars 13; Chicago: The Oriental Institute, forthcoming)..

159 As Astrolabe B puts it, the seventh month is the month when the Anunnaki gods receive the sacrifices of the land, the gates of the Apsû are opened, and *kispu*-offerings are made to Lugaldukuga, Enki and Ninki, the forefathers of Enlil (Tsukimoto, *Totenpflege*, 201: 25-29; 33-38).

160 Labat, *Hémérologies*, 112-113: 22-24 or 170-171: 34-41.

161 Ambos, *Der König im Gefängnis*, 216: A_2 16'-18'; 217: B_1 2-4. For the prayer, see Ambos, *Der König im Gefängnis*, 226-227: B_2 25'-31'.

162 For more on this divinity, see Scurlock 'Nippur's Galileo Problem' in this volume.

163 See Scurlock, 'World Encircling River'.

164 According the omen series *iqqur īpuš*, Tašrītu was the month of the heliacal rising of Libra and the month of the planet Saturn (Tsukimoto, *Totenpflege*, 202: 6).

165 Labat, *Hémérologies*, 168, 169: 8-174, 175: 40 (Assyria); 176, 177:48 -178, 179: 53, 1-34 (Akkad). Another text simply forbids fish on the 2nd and 4th (Labat, *Hémérologies*, 112-113: 20-21, 25-26; 114, 115: 32-116, 117:66 and just about anything you would care to do or eat on the 7th.

to this double seven day to avoid the possibility that the dead might just take somebody living back with them. The complete absence of any activity in the seventh day of the Babylonian seventh month among the common people of the Neo-Assyrian and neo-Babylonian periods also relates to the confrontation between Mercury and Jupiter taking place in that particular month in which they kill each other. With no planet to force them to move, the stars (and time) stand still, and so should you until sunrise of the 8th brings with it the resurrection of the Lords of Time.[166] Already in Mari, the end of the local equivalent of the Babylonian seventh month of Tašrītu was marked not only by a visit of the dead but a sort of joint *akītu*-celebration in which all of the gods of Mari were imagined as leaving for the Netherworld and then returning to their temples to re-establish their relations with the human community.[167]

On this 7th day of the 7th month in Babylonia, a canal is dug in the courtyard of the palace. This is filled with water, and a *makurru*-boat floated on it to facilitate the arrival of the starry participants invoked with sacrifice during the following night. This is fairly certainly the origin of the Rosh Hashanah canal into which sins are ritually to be thrown,[168] but curiously had its ultimate origin in the bowls of water and sailor's throne of Šamši-Adad's original Autumnal ritual.[169]

The Babylonian king of the Neo-Assyrian ritual goes out to a specially constructed structure in the steppeland. It was literally a 'place of imprisonment,' curiously the original meaning in Persian of the term Paradise, and was a walled space with a gate to the East and another gate to the West, leading to the Netherworld. The gate to the East and home is shut fast and the entrance blocked by a *namzītu*-vat. Inside the pallisade was a hut, and both the pallisade and the hut were made of reeds. The king is stripped of his regalia and given a throne special to this ritual.[170] We are seeing here the reed hut of the Nippur Compendium but in the Babylonian version, complete with Bel's *namzītu*-vats.

The king spends the night of the 8th[171] in this hut, hopefully protected by a series of sacrifices to the gods of Nippur and Babylon and various stars including the planet Jupiter, astral manifestation of the god Marduk, the planet Saturn, astral manifestation of the god Anu, the Wagon Star (our Big Dipper), the Arrow Star (Sirius), Orion the three stars of whose belt formed the Bow Star, and the Pleiades.[172] In the morning of the 8th of Tašrītu, before sunrise, the king gets up on a pedestal set out on a reed mat and a linen cloth in the eastern gate of his prison and watched the sun rise over his head to the accompaniment of appropriate sacrifices.[173]

166 It should go without saying that no enlightenment is to be gleaned from statements such as: 'when the ancients looked at their ten fingers or considered the number 10 when counting, their brains instinctively envisaged 10 as three plus a remainder (7). Their brains subconsciously divided reality into 3 and 7. Three was the innate natural number, 'the human number', and the remainder, 10-3, seven, therefore was the non-human number, the number of the 'Other', of the divine' (Cohen, *Festivals and Calendars*, 428). That 'humans', defined as people like 'moderns' (aka Christians) and allegedly Sumerians, who organize their world around the number 3 consider those who organize their world around the number 7 (Zoroastrians, Jews and Muslims) to be 'other' and to operate not by rational thought, of which they are incapable, but by 'instinct' and 'subconcious' processes is well known. We do not need to be reminded of this fact.

167 This will have been the motivation for the sacrificial tour of temples on 27 Mari 9 described in Georges Dossin, 'La Panthéon de Mari,' in *Studia Mariana* (ed. André Parrot et al, Leiden: Brill, 1950), 43-45.

168 For details, see 'Rosh Hashanah,' *Wikipedia* (https://en.wikipedia/Rosh_Hashanah).

169 See above.

170 Ambos, *Der König im Gefängnis*, 216: A_2 19'-25'; 217: B_1 4-17.

171 The night of the 8th will have preceded the day of the 8th by Mesopotamian reckoning.

172 Ambos, *Der König im Gefängnis*, 217-218: B_1 17-52; 219-221: C 1'-23'; A_1 iii 1-21; D1'-8'.

173 Ibid., 221-222: D9'-14'; 222-223: A_2 1-11; B_1 1'-8'. This is not proof that the Assyrian king performed an annual ritual to equate himself with the Sungod (so Eckart Frahm, 'Rising Suns and Falling Stars' in *Experiencing Power--Generating*

While this was going on out in the steppeland, the god Marduk, as we have seen, was himself dressed in a special garment and standing on a pedestal confronting Ninurta at the gates of their respective cellae, echoing the renewal at sunrise of the Chariot of Enlil and the Weapon of Ninurta at Nippur according to the Nippur Compendium as well, presumably, as the Chariot of Nergal at Mari, but with a different hero. What was supposed to be happening, then, was that Marduk was fighting his way home to the upper world. As the king, impersonating Marduk, stood on his pedestal armed with quivers, he and his assisting *āšipu* grasped *mulmullu*-arrows. The king is given his rod of justice, his gold ring crown, his mace, his bow, his ring of power and his stones. He continues to wear or carry all of these except for the bow, the mace and the ring of power, which are taken back by the *ašipu* and put on the royal throne. This done, the king goes home and enters the palace.[174] This return journey corresponds to Marduk's procession of 8 Tašrītu, when he went off to seek the intercession of Gula for which, according to the Nippur Offering List, he had to pay dearly.[175]

All may have been doom and gloom in Babylonia, but in Assyria, the king's annual marriage to Ištar had its denouement after the seven days of abstinence (including an explicit no-sex rule for the 3rd and 6th)[176] in a public celebration in which everybody feasted and renewed their wedding vows in the hopes of engendering a son, a gift from Ištar in which common people could also participate.[177]

It is, in my view, this day that is represented in the famous Aššurbanipal relief of feasting in the garden.[178] The timing is perfect--the campaign against Teumman began in the sixth month. The king is shown feasting with his queen, mandatory, as we have seen for the 8th of the seventh month. She wears the *kulīlu/kilīlu*, the mural crown worn by the goddess Ištar on her wedding day that made her Queen of Heaven,[179] just as Aššurbanipal's wife became Queen of the Empire by virtue of her marriage.

The crown as represented on the relief looks like a miniature set of city walls, complete with towers. This is not what the actual crown looked like, but is a metaphorical representation based on the use of the word for diadem to describe the battlements of a city, you might say its crowning glory. The actual crown is described in cult inventories of the goddess Ištar of Uruk as being a diadem with 30 gold rosettes and 27 precious and [semi-precious] stones as fill.[180]

Authority: Cosmos and Politics in the Ideology of Kingship [ed. Jane A. Hill, et alia; Philadelphia: University of Pennsylvania Museum of Archaeology and Anthropology, 2013], 110-112). It is the Babylonian king who was to perform this ritual, and the purpose of having the Sun rise over him was, as the prayer to the birds at the commencement of the ritual clearly indicates, to ensure his release from his Netherworld prison. 'You, bird of heaven, the creation of [Anu], I, a human being, the creation of N[inmenanna] (Ea), ca[ll] upon you. The fowler caug[ht you] but I saved your [li]fe and let yo[u] see [the light]. You Shamash, protect [my] li[fe]. Just as I g[ave a gift] of life to this bird, give me the gift of my life' (Ambos, *Der König im Gefängnis*, 226-227: B_2 25'-31').

174 Ambos, *Der König im Gefängnis*, 224-226: A_2 12-18; A_1 1'-4'; B_2 9'-20'.

175 See above.

176 Labat, *Hémérologies*, 170, 171: 27-33; 172, 173: 7-17.

177 Ibid., 174, 175: 41-176, 177:46

178 This is mentioned as a comparand for Ramat Rahel feast in Fulton et alia, 'Feasting in Paradis,' 44.

179 Such a crown with its encrusting stones was donned by the goddess Antu, wife of Anu on the 16th of Ayyaru for a joyous visit to the Ekur because the (eternal) Á.SIG$_4$ (sheep) had been put in her lap (OECT 11.69+70 i § 6). The sheep was eternal because this was the part of a woman's dowry that the husband could only use if he replaced it, and it was put into her lap as a sign that she was engaged to be married, which is why she put on the *kilīlu*.

180 Paul Alain Beaulieu, 1999. 'Un inventoire de joaillerie sacrée de l'Eanna d'Uruk,' *Revue d'Assyriologie* 93 (1999), 144: 69.

As luck would have it, we have actual examples of such crowns from the Queen's Tombs at Nimrud. It would seem that, as with Arab women of North Africa to at least my days, it was the custom to bury a woman in her wedding dress. Of the preserved examples, one is comparatively quite plain with just the band and attached rosettes.[181] It was quite flimsy, and probably simply intended to hold on the bridal veil.[182] The other crown is quite spectacular with its circular ribbons of metal holding together rows of gold rosettes inlaid with stones, some of which still have their settings[183] and was, I would argue, the type of crown actually worn by the Queen for the 8th of Tašrītu. I base this on the fact that this crown is also decorated with pomegranates[184] and clusters of grapes glued together with what can only be described as the longest lasting glue ever.[185]

In any case, on Aššurbanipal's relief, we find the royal couple feasting together in the garden. The head of Teumman hanging in the tree overhead, a positive obsession with Western scholars, if taken in an 8 Tašrītu context represents a momento mori that evokes the darker side of this month's activities--the continual propitiation of primordial gods in the absence of the two divinities, Ninurta and Marduk, capable of defeating them, the turn of the year's deciding of destinies as to who would live and who would die. In this context, the head of the dead Elamite king represents those who chose death by opposing Assyria in contrast to the life-choosing tamed lion, a symbolic representation of the Elamite visitors at the edge of the scene.

A special ritual allowed the common people of Babylonia to join with the priests in exorcising this fearful time of year. Offerings in the temple of Gula on the 9th of Tašrītu ensured next years' restoration of Marduk and Ninurta and the continuance of what we call the space-time continuum. For more private needs, common people went out into the steppe on that same day and made offerings to the Goat Star accompanied by a prayer quoted in the hemerologies. It asked, quite simply, for another year of disease-free life.[186] I believe we have found Azazel.

It should by now be obvious that the seventh month of the Hebrew cultic calendar owes not a little to borrowings from Assyria (and later Babylonia in which they lived in exile). These are not, and should not be, seen as adversarial borrowings. Assyria made it possible to believe that brothering could result in unity, and Judah had a brothering problem of the first order, even without the Nile to the Euphrates, in the form of Israel vs. Judah not to mention individual tribes, one of which was allegedly almost wiped out by the others. Besides, as with Šamši-Adad, there was nothing that prevented somebody besides Assyrians taking this daft idea and running with it with or without the acknowledging footnote, even after the Assyrian Empire was nothing but a memory. The modern festivals of Rosh Hashanah and Yom Kippur preserve so many of the elements of the Amorite, Assyrian and Babylonian rituals: the seventh month setting that references original creation, the ram's horns, the fish, the pomegranates, the canal for sins, the *shebat shebatin* taboos, and even Azazael, that it is hard to believe that there is no connection between them.

---

181 This is illustrated in John E. Curtis, et alia (eds.), *New Light on Nimrud: Proceedings of the Nimrud Conference 11-13th March 2002* (London: British Institute For the Study of Iraq, 2008), 106 fig. 14a.

182 Suggestion of MD (Muayyad Said Damerji) apud Dominique Collon, 'Nimrud Treasures: Panel Discussion,' in Curtis, *New Light on Nimrud*, 107.

183 This is illustrated in Curtis et al., *New Light on Nimrud*, pl. 5.

184 MD (Damerji), apud Collon, 'Nimrud Treasures,' 106 wanted these to be opium capsules rather than pomegranates, but the roundness of the seed pods clearly indicates that pomegranates were meant.

185 Donny George Youkhanna, 'Precision Craftsmanship of the Nimrud Gold Material,' in Curtis, *New Light on Nimrud*, 104.

186 Labat, *Hémérologies*, 180, 181: 41-44. By the Neo-Assyrian period, the Goat Star's heliacal rising was no longer at the Autumnal Equinox. However, when it did rise, this could be seen as a positive answer to your prayers.

The Persians Persianized our ritual, transforming the garden into a Persians-only preserve that referenced the time before creation in a uniquely Persian manner that made it difficult for it to be adapted by others for their own use. In the Spring, Nowrūz and in Fall, Mihragān, will have been celebrated in such a garden.[187] It is generally acknowledged that Nowrūz was adopted from the Babylonians.[188] I would argue that so, too, was Mihragān but in a dramatically altered way. A day honoring original creation makes sense in a Babylonian context but the addition of the End of the World certainly does not.[189]

The Persian festival, besides involving a lot of dancing and an opening of vats more in tune with the Ugaritic version of Autumnal Equinox in Araḫšamna,[190] had the king impersonating Mithra complete with a solar crown.[191] The rule that nobody was to look worried or sad and the massive sacrifice of horses, an animal associated in Mesopotamian tradition with Šamaš, god of the sun, mark a time that is aprotropaic as well as celebratory, and the importation of dragon-slaying Faridun from Nowrūz not only serves to link the Autumn with the Spring but also to indicate that some sort of conflict is afoot.[192]

At this time of year, the sun is at a low ebb, in Astrology its nadir,[193] and a Sumerian etiological myth preserved in an Autumnal eclipse ritual still performed in Babylonia in the Seleucid period[194] imagined the Sungod in cahoots with his sister Inanna/Ištar attacking the moon and winding up in defeat, so the Persian complex, if a bit original, still referenced the Babylonian version of the Autumnal equinox ritual with the king ensuring the eventual triumphal comeback of the sun by impersonating Mithra. As for the Assyrian version with its sacred marriage, it is possible that Mithra was imagined as marrying his sister Anahita at Nowrūz and then the king impersonating Mithra married her at Mihragān, thus creating an indirect relationship between the king and Mithra. If so, the only trace in the continuing performance in Iran and India is a custom of eating pomegranates and wishing for 1,000 years of life.[195]

On the other hand, there is the evidence of the Aramaic papyrus in Demotic script found in Egypt in which the Sacred Marriage aspects of the Assyrian ritual are strikingly represented, but in a late antique context that reflects the arrival of Persians and Greeks onto the scene. The authors are apparently people originally moved by Aššurbanipal to Samaria from Raši in the borderlands between Babylonia and Elam (both of whom they hated viscerally). At first, they made themselves at home in Bethel but then were forced to flee to Syene in Egypt where they still dreamed of a return to Raši at the end of time.

Their New Year, celebrated in Tišri, had as its officiant a 'king' who was presumably elected for this purpose. We begin at the gate leading to the courtyard of the New Years' chapel erected for the occasion next to a graveyard. The king stops and recites a blessing; a voice from within

187 Matthew Canepa, 'The Spatial Cosmology of the Iranian Expanse: Transforming Iranian Identity Through Landscape, Architecture and the Built Environment (550 BCE-642 CE),' a lecture given at the Oriental Institute on February 19, 2018.
188 J. Calmard, 'Mihragān,' *The Encyclopaedia of Islam*, New edition, 7 (Leiden: Brill, 1990), 16.
189 For the cosmic associations, see Ibid., 16.
190 JoAnn Scurlock, 'Whirling (*ṣadu*) in the Month of Marzaḫani at Emar or Why did El get, so to speak, drunk as a Lord in KTU 1.114?,' *Ugarit Forschungen* 24 (2013) 285-308.
191 For more details, see Robert Haug 'The Gifts of Mihragān' in this volume. Boyce, apud Calmard, 'Mihragān,' 16 claims that the inspiration was a Babylonian Autumnal festival under the aegis of Šamaš. This is not the case.
192 Calmard, 'Mihragān,' 16-17.
193 Hemerologies mark the 23rd of Tašrītu as the 'eclipse' of the Sungod Šamaš (Labat, *Hémérologies*, 120, 121: 4-5).
194 For details, see JoAnn Scurlock, 'Eclipse of a King: A New Interpretation of the Bull-headed Lyre' in *Proceedings of the 63e RAI Philadelphia July 2016* (forthcoming).
195 EI VII 18.

invites him to enter the courtyard. He enters and washes his hands. The statue of Marah (Nanaya),[196] the Queen of Raši, is brought into the assembly of the gods. The gods rise from their thrones and order that she be seated among them; each of the assembled gods is asked to bless the king. We see here a distant echo of Šamši-Adad's goddess Ištar, surrounded on all sides by a sort of ennead of other divinities among whom is included her brother sungod Šamaš, now 'Our Lord', that is, Nanaya's Mesopotamian husband Nabû who has been syncretized with sungod Apollo/Mithra.[197]

In the Aramaic ritual, sheep are offered in holocaust while 60 singers sing and 60 temple servants burn myrrh and frankincense. The chief god is invited to feast on lamb and to get drunk on wine to the accompaniment of the sweet harp and lyre; force fed-ducks are brought to the table on ivory platters. The king declares: 'Nana, you are my wife' and a priest sings: 'In the bridal chamber, Nana, bring near to me thy lips!'. The king and the goddess keep a vigil outside the bridal chamber with music from a nearby grave to keep them from falling asleep. At the appointed hour, the king invites the goddess to enter the bridal chamber: 'My beloved, enter the door into our house. With my mouth, consort of Our Lord,[198] let me kiss thee!' They enter the perfumed hideaway where the goddess is laid upon an embroidered bedspread. Finally, Nanaya and Ba'al of Heaven[199] exchange blessings and the king promises to rebuild the ruined chapel of Raši.[200]

None of this Persianized ritual was going to play well in Judah, before or after the exile. What I would argue to be the Judean distant descendant of Šamši-Adad's kingship ritual, by contrast, lived on, surviving even the removal of the Judean king from the equation, continuing to symbolically brother widely scattered fellow members of the same religious community across ethnic and even sectarian boundaries to this day.

Appendix

Mari studies are arguably among the most exciting in the field of Assyriology. However, in some areas, particularly religion, they suffer from a fatal combination of too much and too little information. Too much because a surprisingly large number of days seem to have something cultic happening, because the typical sacrifice is not directed to a single divinity but to any number of gods and goddesses whose relevance to the day of sacrifice is more often than not unspecified and because the most popular divinities seem to get sacrifices all the time and not just in their festivals, when these can be discerned. Too little because there are only a few fragmentary ritual texts and no cultic commentaries to help provide the requisite road map, not to mention the fact that you have to deal with a change of dynasty with only a scatter of texts from any given regnal year.

196 Originally a biform of Ištar of Uruk also revered in Elam, she was by this period a super-goddess worshipped all the way to Afghanistan. For details see JoAnn Scurlock, 'Whose Truth and Whose Justice?: The Uruk and Other Late Prophecies Revisited,' in *Orientalism, Assyriology and the Bible* (ed. Steven W. Holloway; Hebrew Bible Monographs 10; Sheffield: Phoenix Press, 2006), 456-462.

197 F. Pomponio, 'Nabû A', *Reallexikon der Assyriologie* 9 (Berlin: Walther de Gruyter, 1998), 23-24.

198 This is her husband, originally the Babylonian god Nabû who was, in late antiquity, equated with Apollo/Mithra. The Babylonian Sungod, Šamaš was Ištar's brother, so you wound up with a brother/sister marriage into the bargain.

199 This is the pair's father, originally the moon god Sîn.

200 Richard C. Steiner, 'The Aramaic Text in Demotic Script: The Liturgy of a New Years' Festival Imported from Bethel to Syene by Exiles from Rash,' *Journal of the American Oriental Society* 111 (1991): 362-363.

Not to despair, since help is actually readily to hand. Since literally Sumerian times, Mari was a part of the Meopotamian umkreis and many of the gods reverenced in Mari are Mesopotamian divinities about whom much more is known both from earlier and from later periods including both rituals and cultic commentaries. It may be presumed that at least some of the cultic activities attested at Mari have ancestors and descendants in Mesopotamia. In order to find these ancestors and descendents, however, it is essential to coordinate the respective calendars. Unfortunately, this has not been done correctly, with results that are, to my mind, somewhere between quite hilarious and enough to make one cry.

In fact, the Mari Calendar as currently understood is a major disaster area, as one who has spent several months rearranging the published references to accord with the actual dates on the tablets can testify. Even armed with an equivalency chart for the alleged sequence of the Šamši-Adad Reichskalander, which is certainly in sync with the accepted sequence of the calendar of Ešnunna,[201] it is difficult to keep references to the same cultic events together, and this not only for non experts. It is also completely unhelpful to present the material by regnal year which predisposes one to assume that any given set of sacrifices is unique to a particular year, with reasons to justify that assumption groped for in the dark. My rearrangement rejoins identical rites on the same calendar dates that have been artificially separated due to one of the entries being dated to the reign of Zimri-lim whose calendar is alleged to have begun in Nisannu and the other to a limmu of Šamši-Adad whose calendar allegedly bizarrely began in Duʾuzu.

When the rearrangement has been completed, it is immedidately apparent that Mari numbered months are somehow out of kilter with their Mesopotamian equivalents. Take, for example, the cultivation of sesame, [202] a Summer crop for which the land was prepared in Mesopotamia in Ayyaru by ploughing the wet areas left over by the flood. In Mari, as we know from a legal case involving a positively modern level of embezzlement, the feed for the plow oxen that our rascally administrator ran off with, preventing their going to work, should have been delivered in Mari 4 (ARM VII 263 ii 14-15). This would make Mari 4 equivalent with Mesopotamian Ayyaru (month 2).

Continuing with Mesopotamian evidence, the early planting (that is, of sesame rather than barley) took place in Duʾuzu (month 4). Sesame ripens in from 2 to 6 months, depending on conditions. There will have been a careful irrigation at the heliacal rising of Sirius (Abu [month 5]) which should have resulted in a harvest in Tašrītu (month 7) or Araḫšamna (month 8), preferably the former, since the latter month needed to be dedicated to ploughing for the all-important barley crop. When the returns for newly harvested sesame at Mari are examined, they range from 23-29 Mari 9 (ARM XXI 135-138). This would make Tašrītu (month 7) equivalent to Mari 9.

In short, both pre-planting plowing and harvest of sesame more than suggest that Mari months are 2 months later in the calendar than their Mesopotamian counterparts. Similarly there is barley on the threshing floors in Mari 3 (ARM VII 263) having been harvested in Magrattum (Mari

201 For Ešnunna see Cohen, *Festivals and Calendars*, 262.

202 For details on this and all agricultural products and their seasons, see Hervé Reculeau, 'Lever d'astres et calendriér agricole à Mari,' in *Recueil d'études à la mémoire d'André Parrot* (Florilegium marianum 6; Mémoires de N.A.B.U 7; Paris: SEPOA, 2002), 523-524.

2) long etymologized as meaning 'threshing floor' or 'grain pile',[203] as would be appropriate for Addaru (month 12). With the current equations, that give Mari 1 as equivalent to Mesopotamian Nisannu (month 1), the Mariots would be harvesting sesame in Kislīmu (month 9), a month after the ground had all been ploughed up for planting barley. They would also be bagging 231 gazelles in Ṭebeṭu (ARM XXI 73), not to mention celebrating festivals throughout the year identical with, but 2 months later than, their Mesopotamian counterparts. I think not.

The only possible conclusion to be drawn from this reassessment is that Šamši-Adad's Reichskalander and that of Zimri-Lim and indeed all Amorite calendars, whatever their names for the months might be, began the year in the Mesopotamian month of Šabaṭu, the Winter Solstice of early Sumerian times (see below pp. 167-68). The following chart gives the equivalencies between the Reichskalander months, the Zimri-Lim months with the allegedly corrected dates now almost universally followed by the Mari scholars in parentheses which add, to my mind, nothing but confusion that consistently gets in the way of proper analysis of the documents.

Nippur Calendar	Mari Calendar	Reichskalander	Eshnunna
11. Šabaṭu	1.Uriaḫum	1. Niqqalu(viii)	1. Niqqalu
12. Addaru	2. Malkanum	2. Magranum(ix)	2. Magrattum
1. Nisannu	3. Laḫḫum	3. Dumuzi[204](x)	3. Elunum
2. Ayyaru	4. Abu	4. Abum[205](xi)	4. Abum
3. Simanu	5. Ḫibirtum	5. Tirum[206](xii)	5. Zibnum
4. Duʾuzu	6. Ḫubur	6. Nikmum[207](i)	6. Nikmum
5. Abu	7. Kinunum	7. Kinunum(ii)	7. Kinunum
6. Ulūlu	8. Dagan	8. Tamḫirum(iii)	8. Tamḫirum
7. Tašrītu	9. Liliatum	9. Nabrum(iv)	9. Nabrum
8. Araḫšamna	10. Belet-biri	10. Mammitum(v)	10. Mammitum
9. Kislīmu	11. Kiskissu	11. Mana(vi)	11. Kiskissu
10. Ṭebētu	12. Eburum	12. Ayyarum(vii)	12. Kinkum

203 See Cohen, *Festivals and Calendars*, 285.

204 The reference is to the heliacal rising (*elûnum*) of astral Dumuzi, the constellation we call Aries, sometimes represented on Mesopotamian cylinder seals as a ram.

205 The name means 'father' and is a reference to 'old man' Enmešarra.

206 The name means 'return' and is a reference to the return of Dumuzi to the Netherworld.

207 The name means 'binding' and is a reference to the 'captivity of the shepherd'

Appendix 2

The Festival of the Great Eštar Day 1:

1 Mari 9 (Tašrītu) textiles and headdresses contributed by the notables and by the *muškēnu* of the districts of Mari, Terqa and Saggaratum when Eštar entered, received by Mukannišum (M 12334+12266 [ARM XXX p. 371-372])

cf. x Mari x garments including one made from ostrich feathers plus shoes issued to Dariš-libur when Eštar (FM XII, p. 22 = M 11909 [ARM XXX p. 236])

cf. x Mari x garments including one red one plus shoes received by [NP] when Eštar (FM XII, p. 22 = M 5199 [ARM XXX, p. 483]; cf. FM XII, p. 227)

1 Mari 9 (Tašrītu): bittersweet *alappanu*-beer for the king and men when (there is) a sacrifice of Eštar in the garden of the king (ARM XII 267-268;[208])

1 Mari 9 (Tašrītu) bittersweet *alappanu*-beer for the meal of the king and men when (there is) a sacrifice in the garden (M 10149 [FM XII, p. 181-182])

1 Mari 9 (Tašrītu) *isqūqum* and *sasqum* flours and *tiktu* (date-drippings) for the table of sweet things (M 10579 [FM XII, p. 158])

1 Mari 9 (Tašrītu) [*isququ*-flour], *tiktu* (date-drippings), *sasqum*-flour for the table of sweet things (ARM XI 112)

cf. x Mari 9 (Tašrītu) superior oil and cedar oil for the bath of Eštar of [...] (RA 69 no. 3[209]; cf. FM XII, p. 160-161)

The Festival of the Great Eštar Day 2:

2 Mari 9 (Tašrītu) women's shoes issued to Inib-šina and Baḫlatum, 2 *ugbabtu*-priestesses and the daughters of the king when Eštar (enters the palace) (M 18164 [ARM XXX p. 332-333])[210]

2 Mari 9 (Tašrītu): bittersweet *alappanu*-beer for the king and men when (there is) a sacrifice of Eštar in the garden of the king (ARM XII 271; cf. FM XII, p. 182)

2 Mari 9 (Tašrītu) bittersweet *alappanu*-beer for the king and men when (there is) a sacrifice of Eštar (ARM XII 270; cf. FM XII, p. 182)

2? Mari 9 (Tašrītu): *buru*-offering, *ḫalluru*-chickpeas and emmer wheat, sourbread for the *zurrayu* jar when Eštar enters the palace (ARM IX 90)

The Festival of the Great Eštar Day 3: ??

208 Cf. Durand and Guichard, 'Les rituels de Mari,' 29.
209 Georges Dossin, 'Tablettes de Mari,' *Revue d'assyriologie* 69 (1975): 26.
210 Cf. Durand and Guichard, 'Les rituels de Mari,' 29.

The Festival of the Great Eštar Day 4:

4 Mari 9 (Tašrītu) [...] when Eštar is made to go up(?) into the shrine of Belet-ekallim; service of Dariš-libur in Mari (ARM XXIV 306[211]).

4 Mari 9 (Tašrītu) x jars for the Babylonian messengers on 29 Mari 8; x jars (expenditures) for the *kannu* [...]; x jars (expenditures) for the *kannu* when Šub-ram, Yarim-lim, Yasmaḫ-Addu, the messengers and *sugagu*'s of the Hanneans sat before the king in the courtyard of the House of the Palms on 4 Mari 9; x jars (expenditures) for the *kannu* when the gods were brought into the temple of Belet-ekallim [on 5 Mari 9] (FM XI 166; cf. FM XII, p. 216)

The Festival of the Great Eštar Day 5:

5 Mari 9 (Tašrītu) superior oil and cedar oil for Nergal when (there is a sacrifice for) his Chariot (ARM VII 28)

5 Mari 9 (Tašrītu) superior oil, superior Mari oil, cypress oil, myrtle oil, *kanaktu* oil and *diqaru*-bowl oil to anoint those attending the throne before the king and the section chiefs (ARM VII 27)

5 Mari 9 (Tašrītu) 15 jars of wine when Kabiya, king of Kaḫat and Yamraṣ-El, man of Abi-ili came when Eštar stayed in the shrine of Bēlet-ekallim (FM XI 57[212])

[5] Mari 9 (Tašrītu) x jars for the Babylonian messengers on 29 Mari 8; x jars (expenditures) for the *kannu* [...]; x jars (expenditures) for the *kannu* when Šub-ram, Yarim-lim, Yasmaḫ-Addu, the messengers and *sugagu*'s of the Hanneans sat before the king in the courtyard of the House of the Palms on 4 Mari 9; x jars (expenditures) for the *kannu* when the gods were brought into the temple of Bēlet-ekallim [on 5 Mari 9] (FM XI 166; cf. FM XII, p. 216)

The Festival of the Great Eštar Day 6:

6 Mari 9 (Tašrītu) a *massilatum* garment for Sametar when Eštar stayed in the shrine of Belet-ekallim (FM XII, p. 22); clothing for Maṣi-ilum; shoes for Zimri-Adad the cook; shoes for Abu-El; expenditures of Dariš-libur at Mari (M 6907 [ARM XXX p. 333][213]).

6 Mari 9 (Tašrītu) distribution of 4 ox carcases in two batches to Iddin-Kakka; distribution of 260 quarters of beef to the cooks (ARM XXI 63)

6 Mari 9 (Tašrītu) 2 jars of wine issued to the lady Bazatum for when Dagan (FM XII, p. 23 = FM XI 13; cf. FM XII, p. 159)

The Festival of the Great Eštar Day 7: Nergal

7 Mari 9 (Tašrītu) oil for the *gibbu* in the temple of Nergal when (there is a sacrifice for the) Chariot (of Nergal) (M 18199 [FM XII, p. 159])

---

211 Ibid.

212 Cf. Durand and Guichard, 'Les rituels de Mari,' 29; Jacquet, *dépenses pour le culte*, p. 183.

213 Cf. Durand and Guichard, 'Les rituels de Mari,' 29.

7 Mari 9 (Tašrītu) cypress oil, *diqaru*-bowl oil, and cedar oil for Nergal when (there is a sacrifice for the) Chariot of Nergal (ARM VII 29)

7 Mari 9 (Tašrītu) *huqu*-bread for king and men when (there is a sacrifice for the) Chariot of Nergal (ARM IX 212 v 1-7; ARM XII 272-273,275[214]).

7 Mari 9 (Tašrītu) bittersweet *alappanu*-beer for king and men when (there is a sacrifice for the) Chariot of Nergal (ARM XII 274[215])

7 Mari 9 (Tašrītu) bittersweet *alappanu*-beer for king and men when (there is a sacrifice for the) Chariot of Nergal (M 10153 [FM XII, p. 183-184])

7 Mari 9 (Tašrītu) bittersweet *alappanu*-beer for king and men when (there is a sacrifice for the) Chariot of Nergal (M 10610 [FM XII, p. 173])[216]

7 Mari 9 the king dedicated a felt cloth and Bini-maraṣ, the miller, to Nergal on the day of the chariot (ARM XXI 328)

The Festival of the Great Eštar Day 7: Eštar

7 Mari 9 (Tašrītu) *ḫašlu*-bread, *mersu*-confection, sour bread, bittersweet *alappanu*-beer, *šipku*-snack (made with) oil, honey and sesame for the king's meal when (there is) a sacrifice of Eštar (ARM IX 131; cf. FM XII, p. 172-173)

7 Mari 9 (Tašrītu) bittersweet *alappanu*-beer for king and men when there is a sacrifice for Eštar in the garden of the king (FM XII, p. 31 = M 10595 [FM XII, p. 172])

7 Mari 9 (Tašrītu) *isququ*-flour, *tiktu* (date-drippings), *ḫašlu*-bread, *sasqu*-flour, and pistachios, totaled as *buru*-offering for the table of sweet things of Eštar (M 10593 [FM XII, p. 171-172])

[7] Mari 9 (Tašrītu) *isqūqu*-flour, *tiktu* (date-drippings), *sasqum*-flour and pistachios for the table of sweet things (M 9994 [FM XII, p. 216])[217]

The Festival of the Great Eštar Day 7: The Feast

7 Mari 9 (Tašrītu) 71 sheep are contributed by the governors and tribal chieftains of the empire (ARM XXIII 222)

undated issue: distribution of 5 cattle heads, 7 cattle shoulders via Bunuma-Addu; 32 sheep shoulders and x cattle shoulders from the garden of the king via Yaṣṣib-Dagan to Ḫaluna-Addu

214 Ibid.

215 Ibid.

216 Jacquet, *dépenses pour le culte*, p. 173 alleges that the chariot of Nergal 'normally' arrived in the palace one week after the sacrifice of Eštar in the garden of the king and that a delay in the performance of the rites of Eštar unique to Zimri-Lim's 7th regnal year caused the two rites to be celebrated simultaneously.

217 Jacquet, *dépenses pour le culte*, p. 216 restores this as [1] Mari 9 to avoid having there being any activity for Eštar on the 7th except in one exceptional year. However, this offering has pistachios as do the sweet things offered on the seventh but not on the first day in any of the references; see above. In any case, the ritual for this festival makes it very clear that not one but two meals were provided for Eštar by the king.

and Yatar-Sumu; x cattle shoulders via Mamma-pilaḫ to feed the craftsmen and Hanneans; 70 fish, 20 baskets of fish via Tamḫiriš-Ḫebat to [...] and Šunuḫruḫalu; 106 fish via Ibni-Sîn and Yaluwumu (ARM XXI 83).

undated issue: distribution of a toal of 51 birds, 248 fish of which 140 fish went to Erib-Sîn and128 fish to Ḫabdu-Kubi; 25 birds went to Erib-Sîn; 26 birds to Ḫabdu-kubi (ARM XXI 92).

The Festival of the Great Eštar Day 8:??

The Festival of the Great Eštar Day 9:

9 Mari 9 (Tašrītu) superior oil for Nergal (MARI 3 12)

The Festival of the Great Eštar Day 10:

10 Mari 9 (Tašrītu) 1 (fattened) ox when there is a *nubattum* (FM XII p. 51 = ARM XXI 15); 1 (fattened) ox when there is a sacrifice in the king's garden (ARM XXI 15-16); 2 (fattened) oxen for the sacrifice of the *ṣirpitu* (barley soup) (ARM XXI 15-16); 1 (fattened) ox and its [...] which they bring from the palace (ARM XXI 15-16); 1 (fattened) ox in the temple of Ea (ARM XXI 15-16); 2 (fattened) oxen when the deities enter the palace (ARM XXI 15-16[218]); 1 (fattened) ox in the temple of Annunitum (ARM XXI 15-16); 1 (fattened) ox for the censer (ARM XXI 15-16); 1 (fattened) ox in the temple of Deritum (ARM XXI 15-16); 1 (fattened) ox when (there is a sacrifice for the) Chariot (of Nergal) (ARM XXI 15-16); x (fattened) oxen in the temple of Annunitu for the second time; total 15 fattened oxen (ARM XXI 15).[219]

218 Cf. Durand and Guichard, 'Les rituels de Mari,' 29

219 Jacquet, *dépenses pour le culte,* p. 201 would like to see this text as a recapitulatif which allows it to be taken apart into individual days that fit his understanding of the ritual activities of this month. The *nubattum*-offering he supposes, as indeed apparently does everybody including the *Assyrian Dictionary of the Oriental Institute of the University of Chicago* N/2 (Chicago: The Oriental Institute, 1980), 207-208, that the reference is to any rite performed in the evening. For this, he is thinking in terms of the pre-nuptial sex mentioned in the ritual as occuring on the evening before day 1 with the sacrifice in the garden being on days 1 and 2 with a single chariot sacrifice for Nergal on day 7. Unfortunately for this argument, there are any number of examples of recapitulatifs in the Mari archives. The monthly recapitulatif usually and the ones for individual festivals always give you the actual day on which a given sacrifice was performed, and some of them even have intermediate subtotals. For this text, we have the ends of lines where such notes appear and we have a total all right but at the end, so all of these sacrifices were made on one day. The remaining lines of the text will have detailed what exactly became of all the bits and pieces from the sacrifices. For what a text of the sort Jacquet is hoping to see here should look like, see especially M 18095 (Jacquet, *dépenses pour le culte,* pp. 212-215). The account of sacrifices may seem like a scarily long list but it pales by comparison with the offerings made on the tour of the temples on the 27th (see above).

Travels through Time

Time, Travel, and the Homesick Hero

Louise M. Pryke

'The marvel of a Home is not that it shelters and warms a man, nor that he owns its walls. It comes from those layers of sweetness which it gradually stores up in us. May it form, deep in our hearts, that obscure range of mountains from which springs the source of our dreams.'
- Wind, Sand and Stars, *Antoine de Saint-Exupéry, 1939.*

Introduction

The concept of 'home' is at the heart of many ancient heroic narratives. 'Home' provides a geographical point of origin from which to set out on a long journey, a familiar counterpoint to measure against foreign lands and characters, and forms an important part of the hero's identity.[1] 'Home' may also contribute to narrative structure, with the return homewards often signaling a journey's end, and the closing arc of a story.

For the heroes of epic narrative in the ancient world, travelling far from home provided opportunities for the accumulation of fortune and glory. Modern scholarship generally accepts that travel in ancient epics was not solely measured in distance and time, but frequently involved more esoteric elements. Voyages may present opportunities for heroes to gain more intimate access to deities or for the development of wisdom.[2] Yet, along with the possibility for divine connection, there was the potential for the journeying hero to become disconnected from home and community. This paper explores the meaning of time in heroic narratives, and the theme of disconnection, through the lens of the motif of the 'homesick hero.' In epic literature, legendary characters such as Gilgamesh, Enkidu, Etana, and Lugalbanda undertake journeys of varying durations, often involving foreign environments that contain natural and supernatural elements. While travel is integral to many heroic narratives, time spent away from home is not presented in a wholly positive light. For Enkidu and Lugalbanda particularly, time away from home results in homesickness and a longing for familiar places and companions.

Through evaluating the impact of time and place on the hero's state of mind, the connection between time and travel in the narratives of Mesopotamian heroes is explored. Lengths of time and changing geographical environments are used in narratives such as *Gilgamesh* to illustrate the epic nature of the hero's journey, with their related risks and drawbacks. The experience of homesickness in the heroic protagonists of epic is presented in a complex manner. Homesickness for heroes involves more than geographical and temporal displacement, but is deeply connected to issues of humanity, mortality, identity, religion, and community.

1 'Home' in this chapter refers to the familiar center of the hero's household or city. The 'home' in Mesopotamia was a complex spatial zone (Klaas R. Veenhof (ed.), *Houses and Households in Ancient Mesopotamia: Papers Read at the 40th Rencontre Assyriologique Internationale, Leiden, July 5–8, 1993* [Publications de l'Institut historique-archélogique néerlandaise de Stamboul 79; Istanbul: Nederlands Historisch-Archaeologisch Instituut in het Nabije Oosten, 1996]). The household formed the basic socio-economic unit, and a home could provide physical shelter while also serving as the location for domestic activities and religious observance.

2 Karljürgen G. Feuerherm, 'Have Horn, Will Travel: The Journeys of Mesopotamian Deities,' in *Travel and Religion in Antiquity* (ed. Philip A. Harland; Waterloo: Wilfrid Laurier University Press, 2011), 83–100.

Time, travel, and heroes

The term 'epic literature' is used here in awareness of the problematic nature of utilizing modern terminology to discuss ancient Mesopotamian literary 'genres.'[3] 'Ancient epic' is a much-debated term, particularly due to the great variety among literature of ancient cultures across different historical and geographical settings. Even when the field is narrowed to specifically consider Mesopotamian 'epic,' the situation remains abstruse. The debate over the definition of 'Mesopotamian epic' has been well encapsulated by recent scholarship.[4] Noegel offers a content-based definition of Mesopotamian epic, stating that he will regard as epic 'all poetic narratives that praise the accomplishments of a heroic figure of history or tradition:' this definition will be adopted here.[5]

Considering depictions of time and travel in Mesopotamian epic literature is a complicated topic. The two concepts are so closely related that they are almost irretrievably enmeshed. Time and travel have distinct significance for the heroic protagonists of epic literature. Heroes, like kings, inhabit a special role that sits in-between the human and divine spheres.[6] The ability to explore and surpass the extremes of human experience, and to reach toward other worlds, sets heroes apart from the rest of humanity. At times, this liminal quality of heroes is expressed through the fantastical qualities of their epic journeys, and the association of socially elite characters with long journeys in turn contributes to the heroic image of epic travel.[7]

Of all the journeys in epic literature, it is perhaps the boundary between humanity and divinity in epic literature that is most frequently explored, and indeed tested, by its heroes. In considering this boundary, the concept of time becomes particularly important. As noted in the *Epic of Gilgamesh*, mortal life inhabits a relatively brief and limited, if unpredictable, span of time (*Gilgamesh* SBV X.319–322).[8] Comparatively, deities may exist forever. While they are often,

---

3 The influential work of H.L.J. Vanstiphout provides a useful overview of generic analyses of Mesopotamian literature. From his 'trail-blazing' paper from 1986, Vanstiphout has built an oeuvre demonstrating the plausibility of fitting styles of Mesopotamian literature into 'genres,' although these genres are still a source of scholarly debate. See notably H.L.J. Vanstiphout, 'Some Thoughts on Genre in Mesopotamian Literature,' in *Keilschriftliche Literaturen. Ausgewählte Vorträge der XXXII Rencontre Assyriologique Internationale, Münster, 8-12.7.85* (Berliner Beiträge zum Vorderen Orient 6; ed. Karl Hecker and Walther Sommerfeld; Berlin: Dietrich Reimer Verlag, 1986), 55–60; H.L.J. Vanstiphout 'I Can Put Anything in its Right Place: Generic and Typological Studies as Strategies for the Analysis and Evaluation of Mankind's Oldest Literature,' in *Aspects of Genre and Type in Pre-Modern Literary Cultures* (ed. H.L.J. Vanstiphout and B. Roest; Groningen: Styx Publications, 1999), 79–101; H.L.J. Vanstiphout 'The Use(s) of Genre in Mesopotamian Literature. An Afterthought,' *Archív Orientální* 67.4 (Jan 1, 1999): 703–717; Jerrold S. Cooper, 'Genre, Gender, and the Sumerian Lamentation,' *Journal of Cuneiform Studies* 58 (2006): 39.

4 Scott Noegel, 'Mesopotamian Epic,' in *A Companion to Ancient Epic* (ed. John Miles Foley; Malden: Blackwell, 2005), 233–45; Jack M. Sasson, 'Comparative Observations on the Near Eastern Epic Traditions,' in *A Companion to Ancient Epic* (ed. John Miles Foley; Malden: Blackwell, 2005) 215–32. For questions over the 'genre' relating particularly to the *Epic of Gilgamesh*, see Andrew George, 'The Epic of Gilgamesh: Thoughts on Genre and Meaning,' in *Gilgamesh and the World of Assyria. Proceedings of the Conference Held at the Mandelbaum House, the University of Sydney, 21-23 July 2004* (Ancient Near Eastern Studies Supplement 21; ed. Joseph Azize and Noel Weeks; Leuven: Peeters, 2007) 37–66.

5 Noegel, *Mesopotamian epic*, 233. The term 'hero' can be considered similarly problematic, as it reflects different meanings in different periods. Jean-Marie Durand, Thomas Römer and Michael Langlois (eds.), *Le jeune héros. Recherches sur la formation et la diffusion d'un thème littéraire au Proche-Orient ancient* (Orbis Biblicus et Orientalis 250; Fribourg: Academic Press, Göttingen: Vandenhoeck & Ruprecht, 2011).

6 Cristiano Grottanelli and Pietro Mander, 'Kingship: Kingship in the Ancient Mediterranean World,' in *Encyclopedia of Religion, 2nd Edition* (ed. Linsay Jones; Detroit: Macmillan Reference USA, 2005), 5163.

7 Mary R. Bachvarova. *From Hittite to Homer: the Anatolian Background of Ancient Greek Epic* (Cambridge: Cambridge University Press, 2016) 209–210.

8 Translations of *The Gilgamesh Epic* are taken from Andrew R. George, *The Babylonian Gilgamesh Epic* (Oxford: Oxford University Press, 2003).

but by no means always, depicted with anthropomorphic characteristics, their lives span an unlimited amount of time.[9]

Time in Mesopotamian literature is a 'subjective notion.'[10] The link between time and travel in Mesopotamian epic is sometimes literal, with time being one method employed by authors to measure distances travelled. On other occasions, time is more esoterically significant.[11] Time and travel are subtly related to concepts of mortality and permanency. Human achievements such as building works, civilization, and 'great deeds' provide a means of reaching beyond the normal limitations of the mortal timeframe. As George has noted, while man is mortal, humanity—and, by extension, civilization—is immortal.[12] For the hero, participating in great deeds brings fame, and a chance for their name to live on beyond the limits of their lives. These heroic deeds often involve long and treacherous journeys.

While some heroic activities, such as wrestling cosmic bulls and battling rival rulers are undertaken close to home, the life of a hero is one that seems to necessitate the accumulation of many ancient frequent flyer miles. Travel is a common feature of the heroic oeuvre, and was also associated with deities.[13] The mythic connection of divinities with the acquisition of power likely influenced the perception of the heroic journey in literature, with the hero following in the footsteps of the divine.[14] Although beyond the scope of this brief survey of the topic, it must be noted that the 'homesick hero' motif, while perhaps originating in Mesopotamia, is not a trope that is isolated to cuneiform literature. Heroes such as Odysseus and Achilles are depicted experiencing homesickness, and also spend large amounts of time away from home.[15] Like Mesopotamian heroes, the favor of the divine, and the observance of the rules of hospitality, are important for the successful navigation of the literary world for Classical heroes. In Egyptian 'travel narratives,' such as the *Story of Sinuhe* and the *Shipwrecked Sailor*, the longing for home creates a dichotomy in the narrative which is useful for exploring the nature of familiarity and foreignness.[16] The significance of these concepts is considered in greater detail below.

---

9 Tallay Ornan, 'In the Likeness of Man: Reflections on the Anthropomorphic Perception of the Divine in Mesopotamian Art,' in *What is a God? Anthropomorphic and Non-Anthropomorphic Aspects of Deity in Ancient Mesopotamia* (ed. Barbara Nevling Porter; Winona Lake: Eisenbrauns, 2009), 93–152.

10 Krzysztof Ulanowski, 'Divine or Human Creation of Time? The Issue of Time as a Factor Determining the Relationship of Man to God,' in *Time and History in the Ancient Near East. Proceedings of the 56th RAI* (ed. Lluís Feliu et al.,; Winona Lake: Eisenbrauns, 2013), 354.

11 Vanstiphout observes the difference between 'natural' and 'supernatural' time in Sumerian epics. He further notes the connection between time and periodization in the literature, with stories often set in a definite period, while showing 'no sense of time perspective.' (H.L.J. Vanstiphout, *Epics of the Sumerian Kings: The Matter of Aratta* [WAW 20; Atlanta: Society of Biblical Literature, 2003], 4–5.

12 George, *The Babylonian Gilgamesh*, 533.

13 Charles Penglase, *Greek Myths and Mesopotamia: Parallels and Influence in the Homeric Hymns and Hesiod*. (London: Routledge, 1994).

14 Kristian Kristiansen and Thomas B. Larsson, *The Rise of Bronze Age Society: Travels, Transmissions, and Transformations* (New York: Cambridge University Press, 2005), 44.

15 Gary S. Meltzer, *Euripides and the Poetics of Nostalgia* (Cambridge: Cambridge University Press, 2006) 33–35. In modern scholarship, Odysseus is frequently associated with nostalgia and Achilles with the emotion of anger. However, Achilles also experiences homesickness (Jan Söffner, 'On Nostalgia [and Homer),' in *Habitus in Habitat I, Emotion and Motion* (ed. Sabine Flach, Daniel Margulies, and Jan Söffner; Bern: Peter Lang AG, 2010), 81–96. During Odysseus's *katabasis*, Achilles expresses a sense of temporal nostalgia as he longs to be alive again, and he experiences a 'disconnect' from his old companions. Achilles is not fixed to a single, geographical nostalgia, rather, he pines for 'earth,' as opposed to the underworld. This temporal nostalgia is juxtaposed with Odysseus's geographical nostalgia (Homer, *The Odyssey: Books 1-12* [trans. A. T. Murray; Harvard University Press—Loeb Classical Library, 1995], pp. 434–37, with reference to lines 11.480–504). My thanks to Marija Elektra Rodriguez for her helpful discussions around 'νόστος.'

16 W. Hutton, 'Travel Writing in the Ancient Mediterranean,' in *The Routledge Companion to Travel Writing* (ed. Carl Thompson; Abingdon: Routledge, 2015).

The experience of travelling away from home may take a hero to a supernatural destination where they can reach beyond the limits of the mortal experience and timespan. In the myth of *Adapa and the South Wind*, Adapa is offered the food and water of life while on a fence-mending visit to Anu in heaven. His rejection of both means that immortality remains beyond his reach, and, perhaps, the reach of the rest of humanity.[17] As Shlomo notes, Adapa's journey allows for the consideration of the distinction between humans and deities, particularly in terms of immortality and wisdom.[18] While visiting the Flood survivor at the end of his journey leads Gilgamesh to accept the finitude of human life, he is offered a rejuvenating plant that would allow him to relive his youth (*Gilgamesh* SBV XI: 278–285). Like Adapa, the hero is unable to capitalize on the opportunity, perhaps reflecting the seriousness with which limitations of mortality are presented in ancient literature. The hero Etana is also able to reach a supernatural destination in the course of an epic narrative when he flies up to heaven.[19] Despite his initial fear in transcending the limits of the earthly realm, it seems probable, despite the broken nature of the text, that his fantastical journey results in the gain of a divine cure for his lack of children.[20]

The hero's means and pace of travel may also reflect their liminality. Gilgamesh and Enkidu travel from the hero's home city of Uruk to the Forest of the Cedars in record speed. As noted by George, if the expression of time is taken literally, the heroes undertake a journey in fifteen days that would ordinarily take seven and a half months.[21] Lugalbanda shows a similar speediness, considered further below. For Etana, the mode of travel to heaven, in flying on the back of an eagle, involves a mixture of divine and animal elements, reflecting the hero's unique position among humanity, as well as the risk involved in attempts to stretch the limits of humanity.

Despite their abnormal capacity for travel, heroes are not immune from the dangers, drudgery, and despair that may accompany a long journey, and the extended absence from home and community. Gilgamesh's journey leads him beyond the path of the sun, and, at this stage of his travels when he is furthest from home, the physical appearance of the notably handsome hero is said to have suffered. As he himself notes to Utanapishti:

> *'Why should my cheeks not be hollow, my face not sunken,*
> *My mood not wretched, my features not wasted?*
> *Should there not be sorrow in my heart,*
> *And my face be not like one who has travelled a distant road?*
> *Should not my face be burnt by frost and sunshine,*
> *And should I not roam the wild got up like a lion?'* (SBV X: 220-225)

Later in the same tablet, Gilgamesh laments the sleeplessness that has accompanied his travels (a sleeplessness that the audience knows will very soon be remedied!):

---

17 Although Liverani makes a good case against this interpretation (Mario Liverani, *Myth and Politics in Ancient Near Eastern Historiography* [ed. Zainab Bahrani and Marc Van De Mieroop; Ithaca; Cornell University Press, 2004], 17–19).

18 Izre'el Shlomo, *Adapa and the South Wind: Language has the Power of Life and Death* (Mesopotamian Civilizations 10; Winona Lake, IN: Eisenbrauns, 2001), 123.

19 The similarities in the journeys of Adapa, Gilgamesh, and Etana are noted by Haul, who observes the unsuccessful quests of the two former heroes raises questions over the success of the latter (Michael Haul, *Das Etana-Epos. Ein Mythos von der Himmelfahrt des Königs von Kiš* [Göttinger Arbeitshefte zur altorientalischen Literatur; Göttingen: Seminar für Keilschriftforschung, 2000], 17–18).

20 Abraham Winitzer, 'Etana in Eden: New Light on the Mesopotamian and Biblical Tales in Their Semitic Context,' *Journal of the American Oriental Society* 133.3 (2013) 445.

21 George, *The Babylonian Gilgamesh Epic*, 463. Swiftness is not, however, entirely reserved for legendary heroes, as can be seen in the speediness of the messenger in the myth of Enmerkar's battle with the ruler of Aratta, and the supernatural speed of Shulgi presented in a royal hymn, also noted in Bachvarova, *From Hittite to Homer*, 210.

'...I went journeying through all the lands,
I passed time and again over arduous mountains,
And I crossed time and again all the seas.
My face did not have enough of sweet sleep,
I scourged myself by going sleepless ...' (SBV X: 251-255).

Gilgamesh further describes the physical toll taken by his travels on his muscles and clothing. The change in Gilgamesh's appearance creates a potent contrast with his earlier physical perfection in Tablet I. The transformation of the king's corporeal state reflects developments in his characterization, as bodily beauty and vitality were integral components of royal imagery.[22] Yet, the epic traveler must face more than a physical cost for his voyages—there is frequently an emotional toll as well, which can be seen in the hero's sense of isolation and homesickness.

The Homesick Hero

The emotional register of nostalgia is a common feature of ancient mythology, and especially prevalent in narratives of return.[23] This chapter considers how the familiar mythological motif of the longing for home is presented in Mesopotamian epic narrative. Does a longing for another, familiar place always relate to a geographical entity, or can 'homesickness' be connected to the desire to return to a particular time, or the company of loved ones? The most plausible answer would be that homesickness for Mesopotamian heroes is, like the concept of time, both a literal and esoteric concept.

It must be stressed that the term 'homesick' describes a modern emotional state which would have been unknown to an ancient Mesopotamian audience, due to cultural and contextual differences, as well as differences in the perception of travel and time. The term 'homesick' in this paper is used as a short-hand to describe a condition where the current locus of the hero is contrasted negatively with a previously inhabited space and/or time—evoking Derrida's 'absent origin' and 'reassuring foundation.'[24] The hero is separated from this 'other' locus and is aware of the disconnection between the present and past state. Homesickness places an emphasis on the hero's desire to travel *towards* rather than *away from* home—either desiring to return to a geographical place, or to a previous time and state. The yearning for home creates a dichotomy between the two points of reference noted by Montiglio as allowing for the conceptualization of travel: the departure, and the destination.[25] The definition of 'homesickness' here is admittedly very broad, in keeping with the frequently nebulous expression of 'homesickness' in epic literature. The idea of returning to a 'simpler time,' an idea commonly known in the modern day, may be classified as 'homesickness' in this definition, as the hero can often be seen to harbor the wish to return to their childhood home or state. The concept of homesickness in Mesopotamian literature is closely related to issues of identity. To explore these ideas further, we now look at the case of Lugalbanda.

22 Beate Pongratz-Leisten, *Religion and Ideology in Assyria* (Studies in Ancient Near Eastern Records 6; Berlin: de Gruyter, 2015), 219–221.

23 Peter Toohey, 'Roman epic,' in *The Cambridge Companion to the Epic* (ed. Catherine Bate; Cambridge: Cambridge University Press, 2010), 31–54.

24 Jacques Derrida, *Writing and Difference*, translated by Alan Bass (Chicago: University of Chicago Press, 1978) 292. Derrida's notion of the dream to recover the 'absent origin' has been used to describe nostalgic tropes in Greek drama (Meltzer, *Euripides*, 35).

25 Silvia Montiglio, *Wandering in Ancient Greek Culture* (Chicago: The University of Chicago Press, 2005) 2.

Lugalbanda

The heroic literary figure Lugalbanda is the protagonist of two Sumerian compositions which involved him travelling a great distance, and spending long periods of time away from home. In both narratives, *Lugalbanda and the Mountain Cave*, and *Lugalbanda*, the legendary hero is engaged in important quests, yet in the course of these narratives he is shown to be in something of a hurry to get back home. Literary tradition places Lugalbanda as the father of fellow-hero Gilgamesh, with the historical Sumerian kings, Shulgi and Ur-Nammu, claiming to be his offspring. The question over the historicity of the early kings Enmerkar and Lugalbanda is one which evades a definitive answer, due to a lack of evidence.[26] If the legendary Lugalbanda is based on an historical person, it is thought that his reign would have occurred around 2800-2700 BCE.[27]

The heroic literary figure of Lugalbanda is the recipient of divine healing in *Lugalbanda and the Mountain Cave* (ETCSL 1.8.2.1). The prayers of Lugalbanda are shown to be effective in gaining the support of the primary deities Utu, Inanna, and Nanna. In Lugalbanda's prayers, he touchingly refers to a seemingly universal and timeless human experience—the ability of illness to inspire profound homesickness. In the storyline of the epic, Lugalbanda first offers a prayer to Utu, for restored health, and describes the unhappiness of his situation in being unwell, alone, and away from home:

> *'Utu, you have let me come up into the mountains in the company of my brothers. In the mountain cave, the most dreadful spot on earth, let me be ill no longer! Here where there is no mother, there is no father, there is no acquaintance, no one whom I value, my mother is not here to say 'Alas, my child!' My brother is not here to say 'Alas, my brother!' My mother's neighbor who enters our house is not here to weep over me.'*[28]

Lugalbanda draws a negative comparison between his current location, 'the most dreadful spot on earth,' and his home where his loved ones would be present to care for him. Prior to leaving him, Lugalbanda's travelling companions attempt to make him comfortable and look after him. In their absence, he longs particularly for those left behind at home. In this way, the company of travelling companions seems to provide a kind of temporary sentimental salve, if not a cure, for the condition of heroic homesickness.

As Lugalbanda prays to the goddess Inanna that he could be with his loved ones, and at home, the king also makes it clear that it is not the grandeur of his home that causes his longing, but rather its familiarity:

> *'Inana, if only this were my home, if only this were my city! If only this were Kulaba, where my mother bore me ...! Even if it were to me as the waste land to a snake! If it were to me as a crack in the ground to a scorpion!'* ETCSL 1.8.2.1

26 Piotr Michalowski, 'Maybe Epic: The Origins and Reception of Sumerian Heroic Poetry,' in *Epic and History* (ed. D. Konstan and K. A. Raaflaub; Oxford: Wiley-Blackwell, 2010) 7–25.

27 Noegel, *Mesopotamian Epic*, 235.

28 Translations of Sumerian epics here are taken from Jeremy A. Black, G. Cunningham, J. Ebeling, E. Flückiger-Hawker, E. Robson, J. Taylor, and G. Zólyomi, *The Electronic Text Corpus of Sumerian Literature* (http://etcsl.orinst.ox.ac.uk/), Oxford 1998–2006.

In this prayer, the author/s of the composition show awareness of the subjectivity surrounding concepts of 'home.' For the snake and the scorpion, there exists a familiar place of belonging, but this place is different for the two creatures, and personal in nature—a 'crack in the ground' holds a different sentimental meaning for a scorpion than it would for a snake.[29] The scorpion analogy holds particularly pleasing connotations of the ability of one's home to provide shelter, but is also the place where one may best 'fit.'

In the course of the narrative, each of Lugalbanda's three prayers have a slightly different emphasis. In his prayer to Utu, he asks for restored health and expresses his desire to be at home. In his prayer to Inanna, he again emphasizes his homesickness. In the prayer to Suen, Lugalbanda requests justice be granted. As the deity primarily identified with love and social connectedness, Inanna is particularly well placed to answer this lonely prayer.[30] Unlike Utu, who sends divine encouragement to Lugalbanda, or Nanna, who gives the king strength, Inanna helps Lugalbanda to go to sleep, and *'(envelopes) him with heart's joy as if with a woolen blanket,'* (ETCSL 1.8.2.1). Inanna's provision of peaceful sleep and 'heart's joy' seem especially well-suited to addressing the heartache and restlessness of homesickness. The reference to being enfolded in a 'woolen blanket' has maternal overtones, considering the ancient Mesopotamian connection of women and craftwork, such as spinning and weaving flax and wool.[31]

Through the divine interventions, Lugalbanda soon recovers, and he makes sacrifices of thanksgiving. It is interesting to note that Lugalbanda's homesickness is chronologically preceded in the narrative by his illness and then his separation from his colleagues—not merely a product of geographical distance, but also involving physical and emotional components. Lugalbanda's physical illness prevents him from moving further *away from home* with his companions, and instead he longs to move *homewards* towards his family and community. The 'homesickness' of Lugalbanda is presented in terms of its seriousness alongside his other, physical illnesses, thereby requiring divine intervention. Lugalbanda's physical illness is the cause of much lamentation by his companions:

> *'... His brothers and friends abandoned holy Lugalbanda in the mountain cave; and with repeated tears and moaning, with tears, with lamentation, with grief and weeping, Lugalbanda's older brothers set off into the mountains.'* ETCSL 1.8.2.1.

For the hero's homesickness to be considered alongside his physical illness in the narrative suggests that the sentimental experience of homesickness was viewed as a serious condition by the author/s of the composition.

In the Sumerian epic *Lugalbanda and the Anzud Bird*, the heroic king again travels a long distance away from home, and is once again separated from his companions. The kindness of Lugalbanda is rewarded by the bestowal of supernatural speed, after he rejects offers of wealth, power, and high status. The epic narrative begins by describing Lugalbanda's lonely and isolated position:

29 The use of snakes in the animal imagery of Mesopotamian narrative and art has often been noted. For example Margaret Cool Root, 'Animals in the Art of Ancient Iran,' in *A History of the Animal World in the Ancient Near East* (ed. Billie-Jean Collins; Handbuch der Orientalistik 1:64; Leiden: E.J. Brill, 2002), 173–180. While less frequently considered in scholarship, scorpion symbolism in Mesopotamian literature and art shows great diversity of usage, and scorpions are often linked to royalty. (Louise M. Pryke, *Scorpion* [London: Reaktion Books, 2016]).

30 Louise M. Pryke, *Ishtar* (London: Routledge, 2017).

31 As Black observes: 'Inana is as reassuring as a lover or mother.' (Jeremy Black, *Reading Sumerian Poetry* [Ithaca: Cornell University Press, 1998], 110).

'Lugalbanda lies idle in the mountains, in the faraway places; he has ventured into the Zabu mountains. No mother is with him to offer advice, no father is with him to talk to him. No one is with him whom he knows, whom he values, no confidant is there to talk to him.' ETCSL 1.8.2.2.

The loneliness of Lugalbanda is described both in terms of his geographical distance from home, and the absence of his companions and loved ones. As in the previous narrative, the hero's separation from his mother is presented as a trying experience. Lugalbanda's 'idleness' may be a product of his homesickness—the text does not describe him as physically unwell, in contrast to the previous narrative.

In his isolated position, Lugalbanda shows careful awareness of the necessity to deal kindly with the inhabitants of his new surroundings, in this case, the offspring of the Anzud bird. Through his actions, he earns the gratitude of the bird, which results in a beneficial reciprocity. The deliberate nature of Lugalbanda's efforts to get the bird 'on side' is shown in his reflective monologue, revealing the hero's intended actions:

'When the bird has drunk the beer and is happy, when Anzud has drunk the beer and is happy, he can help me find the place to which the troops of Unug are going, Anzud can put me on the track of my brothers.' ETCSL 1.8.2.2

Vanstiphout has observed Lugalbanda's choice of reward is a product of his desire to return to his community.[32] This interpretation is supported in the text by the hero's immediate use of his new ability to reunite with his comrades, which emphasizes the importance of social connections for the hero. Once again, sentimental concerns for familiar people and places are favored over material wealth and status by the hero.

The reunion with his companions is described tenderly. First Lugalbanda gives a speech explaining his absence, and then his friends embrace him:

'Lugalbanda's brothers and friends consider the words that he has said to them. Exactly as if they were small birds flocking together all day long they embrace him and kiss him. As if he were a gamgam chick sitting in its nest, they feed him and give him drink. They drive away sickness from Lugalbanda the pure.' ETCSL 1.8.2.2

Once again, it would seem the emotional state of loneliness is the 'illness' affecting Lugalbanda, as it is clear that he has received strength and stamina as a gift from the Anzud bird.

Lugalbanda's supernatural ability also gives him improved access to the Mesopotamian deities, and he is advised by Inanna how to manage his enemy, Aratta. In the narratives of Lugalbanda, the hero is temporarily disconnected from home and companions, but this allows for a closer connection with the divine, and new allies. The development of these closer connections is seemingly conditional on Lugalbanda's appropriate religious observance, and his awareness of the reciprocal nature of hospitality.

32 Vanstiphout, *Epics*, 133.

Enkidu

Unlike Lugalbanda, Enkidu is not a king, and he has neither a mother (at least, prior to his adoption by Ninsun) nor a palatial home to miss. Yet, Enkidu still experiences homesickness. The first suggestion of Enkidu's loneliness comes after his civilizing encounter with Shamhat. After this experience, Enkidu tries to return 'homewards' to his herd, but finds he can no longer live among his former animal companions. The physical and emotional distancing of Enkidu from his herd is engineered by Gilgamesh, who predicts the events which will unfold following Enkidu's meeting with Shamhat, saying:

> *'(Enkidu's) herd will be estranged from him, though he grew up in its presence.'* SBV I.166.

The estrangement of Enkidu from his herd is poignant, especially when placed in context with the *Gilgamesh Epic*'s continued focus on his lack of family. Being part of the herd's community, and sharing their water, brought happiness to his heart—a common result of close, positive social connections in Mesopotamian literature. While the narrative makes clear the benefits of the acculturation process—Enkidu's 'wide understanding'—the process is undoubtedly bittersweet, and involves significant loss.[33] The motif of human-animal separation in *Gilgamesh* has been well elucidated by Barron, who notes the difficulties and costs of the transition from wild animal to civilized human.[34] The growing distance between Enkidu and his origins are illuminated by his increasing violence towards animals, in a sharp contrast with his earlier protective role for wild creatures.[35] Enkidu's 'homesickness' shapes his characterization in the epic, and shows an emphasis on questions of identity and considerations involving the importance of family and community.

Enkidu's process of civilization is successful, in that it leads him to Gilgamesh, the man whom he was created to counsel and accompany. Yet, despite the narrative's appreciation of the products of civilization, the acculturation of Enkidu is not presented as entirely beneficial to him, or without sacrifice. Enkidu moves from a state of happy innocence to acquiring reason, and his acculturation seems at times to be unwilling, and emotionally confusing.

When Enkidu discovers he can no longer keep up with the herd, he returns to his new lover. Shamhat presents to Enkidu the religious setting of life in Uruk, and the companionship of Gilgamesh, as alternatives to the ability to 'roam the wild with the animals.' Enkidu accepts these alternatives, due to his loneliness; the text clarifies the young hero's motivations:

> *'(Enkidu's) heart (now) wise was seeking a friend.'* SBV I.214.

Now, estranged from his home, Enkidu seeks companionship.

Despite Enkidu's love for Gilgamesh, his loneliness and pining for his former life in wild are a continued part of his characterization in the narrative, until his death. Indeed, in keeping

33 Ulla S. Westenholz and Christian Felix Westenholz, 'Enkidu—the Noble Savage?' *In Wisdom, Gods and Literature: studies in Assyriology in honour of W. G. Lambert.* (eds. Andrew R. George and Irving L. Finkel; Winona Lake: Eisenbrauns, Inc., (2000) p. 444.

34 P. Barron, 'The Separation of Wild Animal Nature and Human Nature in Gilgamesh: Roots of a Contemporary Theme,' *Papers on Language and Literature* 30.4 (2002): 393.

35 Barron, 'Separation', 390–391. As noted by Barron, Enkidu's violence towards animals causes a dissociation with his former identity, but also contributes to his alienation from Uruk's deities. See Jared Christman, 'The Gilgamesh Complex: The Quest for Death Transcendence and the Killing of Animals,' *Society and Animals* 16 (2008): 302.

with the complexity of the presentation of homesickness in epic narratives, in Tablet III of the Standard Babylonian Version of the *Epic*, Gilgamesh seems to suggest travel and heroic ventures as a cure for Enkidu's sense of being out of place in his new environment of Uruk.

On his deathbed, Enkidu expresses his regrets about his process of civilization, and lays heavy curses on the trapper and Shamhat—the pair he perceives as responsible for his unhappy situation. There is irony to this dramatic scene, as the audience is aware that it is Enkidu's beloved Gilgamesh who has engineered his acculturation, and whose belligerent behavior at the start of the narrative underlies Enkidu's very existence. Again, in this scene, however, Enkidu is presented with the alternatives to his wild life. The solar deity Shamash questions Enkidu's choice to curse Shamhat—seemingly the trapper is fair game for Enkidu's maledictions. In a striking piece of dialogue, Shamash speaks directly to Enkidu from the sky, asking:

> *'Why, O Enkidu, do you keep cursing the harlot Shamhat,*
> *who gave you bread to eat, fit for a god,*
> *gave you beer to drink, fit for a king,*
> *clothed you in a great garment,*
> *and let you have for a comrade the fine Gilgamesh?'* SBV VII. 134–138.

In Shamash's speech, we see the products of civilization, such as clothing and eating bread, conceptually related to drawing Enkidu away from his animal state, and towards divinity and royalty. The love of Gilgamesh, and the hero's capacity to provide Enkidu with appropriate mourning rituals following his death, are the focus of the second half of Shamash's speech:

> *'Now Gilgamesh, your friend and brother,*
> *[will] lay you out on a great bed...*
> *[And] he, after you are gone he will have himself bear the matted hair of mourning,*
> *[he will don] the skin of a lion and go roaming in the wild.'* SBV VII.139–143; 146–147.

Gilgamesh's emotional closeness to Enkidu and the familial connection between them are provided by Shamash as reasons for Enkidu's 'angry heart' to be placated. The reference to Gilgamesh as Enkidu's 'brother' possibly relates to Enkidu's adoption by Ninsun, but more probable is a metaphorical use of the word to express the family-like closeness between the two heroes. The words of Shamash, like those of Shamhat earlier in the narrative, have the immediate effect of changing Enkidu's outlook, and cause him to take action. Instead of cursing Shamhat, after Shamash's speech, he blesses her. The close companionship between Enkidu and Gilgamesh is presented as a good alternative to Enkidu's original community in the wild.

Gilgamesh

As the heroic protagonist of the most famous epic narrative from ancient Mesopotamia, few literary figures can claim to be so widely travelled as Gilgamesh. Yet, despite his evangelizing of epic travels to Enkidu, and his many incredible journeys, Gilgamesh may also be said to show a sense of 'homesickness.'[36]

36 The religious rituals of Gilgamesh's mother, Ninsun, and the intervention of Shamash protect Gilgamesh while journeying far from home (Sabina Franke, 'Fortschreibungsprozesse im/des Gilgamesch-Epos,' in *Identität und Schrift: Fortschreibungsprozesse als Mittel religiöser Identitätsbildung* [ed. Marianne Grohmann; Biblisch-theologische Studien 169; Gottingen: Vandenhoeck & Ruprecht, 2017], 2). Again, this demonstrates the connection between travel safety and religiosity.

After meeting with the Flood survivor, Utanapishtim, Gilgamesh learns that immortality is beyond the reach of humans, and even quasi-divine beings like himself. Utanapishtim offers Gilgamesh the *amurdinnu*, or heartbeat plant, as a kind of 'runner's up' prize in the place of immortality. While it won't make the hero live forever, he will be able to return to his youth. Gilgamesh's search and acquisition of the plant can be considered as the expression of longing to return homewards to the time of his youth. Prior to discovering the truth of his own mortality, Gilgamesh continually pushes onwards and often moves away from home. However, in contrast to his typical forward momentum, the plant's very nature is regressive.[37]

The placement of this episode involving the *amurdinnu* plant in the structure of the narrative is particularly interesting—it occurs just after Gilgamesh has accepted his mortality, and just prior to his turning homewards. Considered in this light, 'homesickness' is a particularly human condition in Mesopotamian epic literature, and one powerful enough to make even the most intrepid and prodigious traveller set their sights on a return voyage. For Gilgamesh, absence has made the heart grow fonder: the event of the young king's 'home-coming' is emphasised by his praise of the walls of Uruk. Gilgamesh's speech firmly settles him back in his original geographic location, while demonstrating his renewed appreciation for his home.[38]

Conclusion

Mesopotamian heroes undertake long journeys and take part in great deeds, resulting in fame and glory that often surpasses the limited mortal timeframe. However, these journeys can result in a disconnection from home, and the emotional experience of homesickness. The longing for home in the characterisation of the heroes of Mesopotamian epic presents a truly universal experience in some of the world's most ancient literature, and performs an important narrative function. Recent studies on emotions and affect in ancient narrative emphasize the concept of emotions not only as fixed and internalized expressions, but as forces which can be transformative and create action.[39] As much as the benefits offered by far flung places can improve the hero's social and material standing, the emotional experience of homesickness plays a significant role in drawing the hero back homewards again at the end of his adventures.

After all, even for a Mesopotamian hero, there's no place like home.

37 As Forest notes, Gilgamesh manages to 'transform' even without the aid of the plant. Jean-Daniel Forest, 'L' Epopée de Gilgameš, ses origins et sa postérité,' in *Gilgameš and the World of Assyria: Proceedings of the Conference Held at the Mandelbaum House, The University of Sydney, 21-23 July 2004* (eds. Noel Weeks and Joseph Azize; Ancient Near Eastern Studies Supplement 21; Leuven: Peeters, 2007), 32.

38 The return to the walls of Uruk signal the completion of the king's journey, and contributes to the development of his character. The scene in Tablet XI on the walls of Uruk adds meaning to Gilgamesh's long journey, and his search for immortality. See Annette Zgoll, 'monumentum aere perennius—Mauerring und Ringkomposition im Gilgamesh-Epos,' in *Von Göttern und Menschen. Beiträge zur Literatur und Geschichte des Alten Orients. Festschrift für Brigitte Groneberg* (eds. Dahlia Shehata, *et al.*; Cuneiform Monographs 41; Leiden: Brill, 2010), 443–470.

39 Well encapsulated by Sophus Helle, *Emotions in Gilgamesh: Desire, Grief and Identity in the Standard Babylonian Epic of Gilgamesh*. MA Thesis, University of Copenhagen, 2016.

The Older the Better: The Critical Conception of Lateness in Song China

Xiaoshan Yang

Introduction

There is a long-standing idea in Chinese literary-critical tradition that a great writer achieves his best late in life. In this paper, I situate the genesis of this idea of lateness in the cultural process in the Song period (960–1279) through which Du Fu 杜甫 (712–770) was canonized as the greatest of all Chinese poets. I begin with a brief review of his canonization from its inception in the ninth century to its completion in the eleventh. I then examine a critical practice that was rooted in and contributed to this canonization, i.e., that of periodizing his poetry, defining the salient feature of each period, and singling out his late works as the culmination of his genius. I illustrate how this model of hierarchical periodization was then applied in the criticisms of major Song poets. In the third section of this paper, through a case study of the critical evaluations of Wang Anshi 王安石 (1021–1086), I discuss the simultaneous elevation of lateness and repudiation of juvenility. I conclude with some cross-cultural comparative reflections.

Canonizing Du Fu

Du Fu's critical ascension to the apex of the Chinese poetic tradition started about four decades after his death.[1] In a tombstone epitaph composed in 813 at the request of Du Fu's grandson, Yuan Zhen 元稹 (779–831) defined the poet's genius in terms of its unprecedented thematic, formal, and stylistic comprehensiveness. He also initiated the comparative ranking of Li Bai 李白 (701–762) and Du Fu that would obsess critics for ages to come: Li Bai may be on equal footing with Du Fu in terms of his heroic mannerism and unrestrained abandon and in such genres as *yuefu* poems and ballads, but he does not come anywhere close to Du Fu in regard to the mastery of long regulated verse poems 排律, characterized, among other things, by extended narration, skillful arrangement of melody and rhyme, and fitting parallelism.[2]

Yuan Zhen's establishment of the superiority of Du Fu over Li Bai on purely technical grounds has been the subject of much criticism.[3] He was not, however, unaware of the moral weight of Du Fu's poetry. In a letter to Bai Juyi 白居易 (772–846) in 815, he reminisced about the impact of his personal encounter with Du Fu's poetry: When he was a young man, he was inspired by Chen Ziang's 陳子昂 (661–702) 'Moved by Experiences' 感遇 series. Later, he obtained several

1 Scholarship in Chinese is legion on of Du Fu's canonization. For discussions in English, see Shan Chou, *Reconsidering Tu Fu: Literary Greatness and Cultural Context* (Cambridge and New York: Cambridge University Press, 1995), 1–59; Charles Hartman, 'The Tang Poet Du Fu and the Song Dynasty Literati,' *Chinese Literature: Essays, Articles, Reviews* (*CLEAR*) 30 (Dec., 2008): 43–74; Hao Ji, *The Reception of Du Fu (712-770) and His Poetry in Imperial China* (Leiden: Brill, 2017).

2 Yuan Zhen, 'Tang gu Gongbu yuanwailang Du jun muximing' 唐故工部員外郎杜君墓係銘 (Tombstone Epitaph to Mr. Du, the Late Vice Director in the Ministry of Works of the Tang), *Yuan Zhen ji* 元稹集 (Collected Works of Yuan Zhen), 56.691.

3 See, for example, Yuan Haowen 元好問 (1190–1257), 'Lun shi sanshi shou' 論詩三十首 (no. 10), *Yuan Haowen shi biannian jiaozhu* 元好問詩編年校注 (The Poems of Yuan Haowen in Chronological Order, Collated and Annotated) (Beijing: Zhonghua shuju, 2011), 1.54; Hu Zhenheng 胡震亨 (1569–1649), *Tangyin guiqian* 唐音癸籤 (Comments on Tang Poetry, the Tenth Series) (Shanghai: Shanghai guji chubanshe, 1981), 6.56.

hundred poems of Du Fu and was impressed by their limitless expansiveness and all-around perfection in addition to their 'allegorical lyricism' 寄興.[4]

In his reply to Yuan Zhen, Bai Juyi foregrounded the social-political function of poetry and found both Li Bai and Du Fu wanting: Li Bai is marvelously talented, but, if one were to look for 'moral symbolism' 比興 in the mode of *Airs of the States and Odes, one would not find it in one out of ten poems. Du Fu, for all his technical superiority over Li Bai, is also a moral lightweight in numeric terms: out of the large corpus of his poems totaling more than a thousand, no more than thirty or forty are pieces of social criticism.*[5]

Yuan Zhen's comparative ranking of Li Bai and Du Fu was targeted in the beginning lines of Han Yu's 韓愈 (768–824) 'Written Playfully to Zhang Ji' 調張籍, composed in 816:[6]

李杜文章在	Li Bo's and Du Fu's writings endure,
光燄萬丈長	a blazing halo ten thousand yards high.
不知群兒愚	I don't understand that pack of fools,
那用故謗傷	why did they malign them so willfully?
蚍蜉撼大樹	Such are ants that would shake a mighty tree,
可笑不自量	absurdly blind to their own capacity.

The poem goes on to describe hyperbolically the overwhelming power of the poetry of Li Bai and Du Fu and whimsically speculate about the ontology of their lifetime frustrations. Near the end of the poem, Han Yu fantasizes about joining the company of the dead poets:

我願生兩翅	My wish is to grow a pair of wings
捕逐出八荒	to hunt them down past the world's confines.[7]

The critical interest in Du Fu's poetry in the second decade of the ninth century was not sustained for long. His influence was rather limited thereafter in the Tang and Five Dynasties periods. The situation remained fundamentally the same in the first half century after the founding of the Song dynasty. None of the three prominent styles of poetry in the early Song took its model from Du Fu. The leading poets in the second half of the tenth century such as Li Fang 李昉 (925–996), Xu Xuan 徐鉉 (916–991), and Wang Yucheng 王禹稱 (954–1001) imitated aspects of Bai Juyi's poetry, hence the characterization of their poems in terms of *Bai ti* 白體 (Bai style). The term *Wan Tang ti* 晚唐體 (late Tang style) was applied to the works of a diverse group of poets—ranging from the Nine Monks 九僧 and recluses like Lin Pu 林浦 (967–1028) and Wei Ye 魏野 (960–1019) to the high ranking official Kou Zhun 寇準 (961–1023)—who found inspiration in the late Tang poets Jia Dao 賈島 (779–843) and Yao He 姚合 (781?–846). Finally, the densely allusive poetry of Li Shangyin 李商隱 (813?–858) served

4 Yuan Zhen, 'Shushi ji Letian shu' 敘詩寄樂天書 (A Letter to Letian about Poetry), *Yuan Zhen ji, 30.406.*

5 Bai Juyi, 'Yu Yuan Jiu shu' 與元九書 (A Letter to Yuanjiu), *Bai Juyi ji jianjiao* 白居易集箋校 (Collected Works of Bai Juyi, Collated and Annotated) (ed. Zhu Jincheng 朱金城; Shanghai: Shanghai guji chubanshe, 1988), 45.2791. One major textual variant has 'thirteen for fourteen' instead of 'thirty or forty.'

6 The connection between Han Yu's poem and Yuan Zhen's comment was first made in Wei Tai, *Lin Han yinju shihua jiaozhu* 臨漢隱居詩話校注 (Remarks on Poetry by the Recluse upon the Han River, Collated and Annotated) (ed. Chen Yingluan 陳應鸞; Chengdu: Bashu shushe, 2001), 1.21–22.

7 Han Yu, *Han Changli shiji biannian jianzhu* 韓昌黎詩集編年箋注 (Collected Poems of Han Changli in Chronological Order, with Commentaries and Notes), compiled by Fang Shiju 方世舉 (1675–1759) and edited by Hao Runhua 郝潤華 and Ding Junli 丁俊麗 (Beijing: Zhonghua shuju, 2012), 9.517–18; English translation by Stephen Owen, *An Anthology of Chinese Literature: Beginnings to 1911* (New York: Norton, 1996) 487.

as the model for poets of *Xikun ti* 西昆體 (Xikun style).[8] The name derived from the title of *Xikun chouchang ji* 西昆酬唱集 (An Anthology *of Poems Exchanged in the Xikun Library), compiled around 1005 by Yang Yi* 楊憶 (974–1020), one of the seventeen imperial librarians whose works were included in the anthology and a leading man of letters known for his contempt for Du Fu as a 'village pedant' 村夫子.[9]

The drive to canonize Du Fu received renewed momentum in the third decade of the eleventh century, with scholarly efforts to collect and edit his works. In the early 1020s, Su Shunqin 蘇舜欽 (1008–1048) culled three hundred poems from a Du Fu collection that were not included in the extant old collection. Then, sometime between 1035 and 1036, he obtained another collection that contained over eighty poems not found anywhere else. He was confident that these poems could not have been written by anyone else because they all exhibit Du Fu's recognizable and yet inimitable style of 'heroic abandon and plaintive cadence' 豪邁哀頓.[10]

Su Shunqin's confidence is based on a kind of poetic connoisseurship of which the premise is that there is something unique about the stylistic and emotive efficacy of Du Fu's poetry, though different critics would characterize it in different terms. In a 1052 preface to a Du Fu collection that he had just obtained, Wang Anshi mentioned his ability to determine the authenticity of a poem attributed to Du Fu on the basis of the distinctness of the language and conception of Du Fu's poetry.[11] Wang Anshi's greatest contribution to the canonization of Du Fu lies in his poem 'On a Portrait of Du Fu' 杜甫畫像, which transforms the Tang poet into a moral icon as well as a poetic master:

	吾觀少陵詩	When I read Du Fu's poetry,
	為與元氣侔	I feel in touch with the primordial spirit.
	力能排天斡九地	Its power can push heaven and earth around;
4	壯顏毅色不可求	His resolute look can no longer be captured.
	浩蕩八極中	In this vast world extending in eight directions,
	生物豈不稠	What a multitude of creatures there is!
	醜妍巨細千萬殊	The ugly and the beautiful, the grand and the minute: a myriad of differences;
8	竟莫見以何雕鎪	And yet there is no seeing how he carved and engraved.
	惜哉命之窮	What a pity that he was destined to impoverishment:
	顛倒不見收	Never summoned to office after all his tribulations.
	青衫老更斥	Robed in blue, all the more ignored in old age,[12]
12	餓走半九州	He wandered half of the nine domains, starved.
	瘦妻僵前子仆後	His gaunt wife fell first, then his children;

8 The three styles of poetry in the early Song period was first summarized in Fang Hui 方回 (1227–1307), 'Song Luo Shouke shi xu' 送羅壽可詩序 (Preface to a Poem Sent to Luo Shouke), in *Quan Yuan wen* 全元文 (Complete Prose of the Yuan Dynasty), 60 vols. (edited by Liu Xiusheng 李修生 ; Nanjing: Jiangsu guji chubanshe-Fenghuang chubanshe, 1999–2004), 7:209.51.

9 Liu Ban 劉攽 (1023–1089), *Zhongshan shihua* 中山詩話 (Zhongshan's Remarks on Poetry), in *Lidai shihua* 歷代詩話 (Remarks on Poetry from Various Dynasties), comp. He Wenhuan 何文煥 (1732–1809) (Beijing: Zhonghua shuju, 1981), 288.

10 Su Shunqin, 'Ti Du Zimei bieji hou' 題杜子美別集後 (Colophon to the Separate Collection of Du Zimei), in *Quan Song wen* 全宋文 (Complete Prose of the Song Dynasty), eds., Zeng Zaozhuang 曾棗莊, Liu Lin 劉琳 et al., 360 vols. (Shanghai: Shanghai cishu chubanshe, 2006) [hereafter *QSW*], 41:878.72.

11 Wang Anshi, 'Preface to a Second Collection of Du Fu's Poems' 老杜詩後集序, *QSW, 64:1398.273.*

12 Officials of the eighth and ninth grades wore blue robes in the Tang dynasty; see Liu Xu劉昫 (888–974), *Jiu Tang shu* 舊唐書 (Old Tang History) (Beijing: Zhonghua shuju, 1975), 45.1953.

攘攘盜賊森戈矛	Hordes of rowdy rebels raised a forest of spears.
吟哦當此時	Even then, in his chanting and intoning,
不廢朝廷憂	He did not cease in his concern for the court:
常願天子聖	Always wishing the Son of Heaven to be sagely,
大臣各伊周	And great ministers to be like Yi and Zhou.[13]
寧令吾廬獨破受凍死	He'd rather his cottage break and he freeze to death
不忍四海赤子寒颼颼	Than folks of the four seas endure the whooshing cold.[14]
傷屯悼屈止一身	Moaning and whining only about their own wretched misery—
嗟時之人我所羞	Alas, how ashamed I feel about people these days!
所以見公像	That is why when I see your portrait,
再拜涕泗流	I bow again and again with tears streaming down.
惟公之心古亦少	A mind like yours has been rare since antiquity;
願起公死從之游	I wish you could rise from death so that I may follow you around.[15]

The first part of the poem (lines 1–8) encapsulates the critical discourse on the sublimity and range of Du Fu's poetry and the ungraspableness and inimitableness of his art. It is in the second part (lines 9–22) that Wang Anshi adds a new dimension to Du Fu's greatness. The Tang poet now emerges as a moral paragon with his loyalty to the sovereign and concern for the common people, in spite of his political frustrations and personal misfortunes. Du Fu's public-mindedness is set in bold relief against the pettiness of 'people these days' wallowing in their self-pity. Wang Anshi's expression, in the last couplet of the poem, of the desire to join Du Fu's companionship is reminiscent of the ending of Han Yu's poem, but there is a crucial difference: Han Yu's tone is deliberately and decisively fanciful; Wang Anshi, on the hand, is deadly serious.

Wang Anshi's has been singled out as the best poem on Du Fu's portraits.[16] While it is an exaggeration to claim that '[the] view commonly accepted today that [Du Fu] surpassed all other poets of past and present was first expressed by Wang [Anshi],' he certainly played a crucial role in establishing Du Fu's supremacy on moral as well as aesthetic grounds.[17] As Qiu Zhaoao 仇兆鼇 (1638–1717) wrote, the poem hits the mark about Du Fu's character and integrity as well as his poetic talent so that later admirers had nothing more to add.[18] Indeed, after Wang Anshi, no meaningful new ground was broken in the canonization of Du Fu, despite all the progress in the scholarship on various aspects of his life and works.

13 Yi refers to Yi Yin 伊尹; Zhou, to the Duke of Zhou 周公. Both were exemplary ministers of high antiquity.

14 This is a paraphrase of the concluding lines of Du Fu's 'A Song on How My Thatched Roof Was Ruined by the Autumn Wind' 茅屋為秋風所破歌, *The Poetry of Du Fu* 杜甫詩, tr. and ed. Stephen Owen, 6 vols. (Boston: De Gruyter, 2016), 3:42–43.

15 In *Quan Song shi* 全宋詩 (Complete Song Dynasty Poetry), comp. Beijing daxue guwenxian yanjiusuo (Beijing: Beijing daxue chubanshe, 1991–98) [hereafter *QSS], 546.6537.*

16 See Liu Kezhuang 劉克莊 (1187–1269), *Houcun shihua* 後村詩話 (Houcun's Remarks on Poetry) (Beijing: Zhonghua shuju, 1983), *xinji* 新集 (New Collection), 1.152; Hu Zi 胡仔 (fl. 1147–67), *Tiaoxi yuyin conghua* 苕溪漁隱叢話 (Remarks on Poetry by the Fisherman Recluse of the Tiao Stream), ed. Liao Deming 廖德明 (Beijing: Renmin wenxue chubanshe, 1962), *qianji* 前集 (First Collection), 11.72.

17 Yoshikawa Kōjirō, *An Introduction to Sung Poetry* (tr. Burton Watson; Cambridge, Mass.: Harvard University Press, 1967), 92. Yuan Zhen, as we have seen, had already described Du Fu as the greatest poet in terms of the thematic, formal, and stylistic variety of his poetry. Wang Anshi's contribution lies in his elevating Du Fu as a moral paragon.

18 Qiu Zhaoao, *Du shi xiangzhu* 杜詩詳註 (The Poetry of Du Fu, with Detailed Annotations) (Beijing: Zhonghua shuju, 1979), 2268.

Periodizing a Poet's Lifework

In his 1036 preface, Su Shunqin lamented that only twenty *juan* 卷 (scroll) of Du Fu's poems had survived, though his biography mentioned a sixty-*juan collection. Furthermore, the poems had not been edited by scholars: ancient-style*古體and modern-style 近體 poems are mixed together, with no chronological sequence. Once he obtained the complete poems of Du Fu, he planned to separate the ancient-style and recent-style poems and then place them in chronological order within each category.[19] This plan, though never carried out, was hatched on a new editorial principle, which departed from the well-established convention of ordering poems by themes or genres. The chronological model reflected the strong desire to read Du Fu's poems in light of his biography. Such a desire was incited and nourished in part by Du Fu's constant and careful account of his own life.

The appearance of *nianpu* 年譜 (year-by-year biography) in the late eleventh century was prompted by, and in turn provided further momentum to, the biographical reading of Du Fu's poems. In the1080s, Lü Dafang呂大防 (1027–1097) produced the first full-fledged chronological editions of Du Fu's poetry and Han Yu's prose. He also composed a *nianpu* for each and wrote the following colophon (dated December 13, 1084):

> I was troubled by the many errors in [the collections of] Han Yu's prose and Du Fu's poetry, so I emended them. I also composed a *nianpu* for each of them so as to document the dates of their writings and to provide a glimpse into the times during which they wrote. Thus, their praise of the time as well as their lamentations about the world, their hidden concerns as well as their explicit grief all become clearly observable. Their verbal vitality can also be studied, which is sharp in youth, unbridled in prime, and rigorous in old age. One cannot reach such an attainment without subtlety in writing.
>
> 予苦于韓文杜詩之多誤，既讎正之。又各為年譜，以次第其出處之歲月，而略見其為文之時，則其歌時傷世、幽憂切歎之意，粲然可觀。又得以考其辭力，少而銳，壯而肆，老而嚴。非妙于文章，不足以至此。[20]

Lü Dafang not only initiated *nianpu* as a new genre of historiographic writing but also authoritatively defined the double purposes of attaching it to literary collections, i.e., to provide historical background for the texts and to shed light on the evolution of a writer's 'verbal vitality' 辭力. This twofold teleology would be routinely reiterated by later compilers. In a colophon dated 1135, for example, Wen Anli 文安禮 explained that he compiled Liu Zongyuan's 柳宗元 (773–819) *nianpu* 'so that the sources and times of his writings may be known and the development of his verbal vitality may be investigated' 庶可知其出處與夫作文之歲月，得以究其辭力如何也.[21] In 1161, Chen Hui 陳輝 asked his son Chen Ye 陳曄 to compose a *nianpu* of Chen Xiang 陳襄 (1017–1080) to be appended to the latter's literary collection 'so that the chronology of his life and career, his aspiration in prime and integrity in old age, and his verbal vitality in poetry and prose may become clearly discernible' 庶幾生平遊宦歲月之先後，與夫壯志晚節，詩文之辭力，曉然可見.[22]

19 Su Shunqin, 'Ti Du Zimei bieji hou.'

20 Lü Dafang, ‹Du Gongbu Han Wengong nianpu houji› 杜工部韓文公年譜後記 (Colophon to the Chronological Biographies of Du of the Ministry of Works and the Literary Duke Han), *QSW*, 72:1573.209.

21 An Wenli, 'Liu wen nianpu houxu' 柳文年譜後序 (Colophon to the Chronological Biography Attached to the Prose of Liu), *QSW*, 186:4088.115.

22 Chen Ye, colophon to 'Guling xiansheng nianpu' 古靈先生年譜 ('Chronological Biography of Master Gulling),

In Lü Dafang's scheme, the literary career of a writer is periodized in accordance with the three biological phases of youth, prime, and old age, with his 'verbal vitality' distinctly manifested at each phase. In an 1100 letter to Wang Fan 王番, Huang Tingjian 黃庭堅 (1045–1105) advanced a different paradigm as he drew on the examples of Du Fu and Han Yu to advise his friend against an obsessive pursuit of extraordinary poetic diction:

> The penchant for strange phraseology is a fault in writing. You should take Reason as the principle. When you attain to Reason, your words will flow smoothly, and your writing will be outstanding naturally. Look at Du Zimei's [Du Fu] poetry after he arrived at Kuizhou and Han Tuizhi's [Han Yu] prose after he returned to court from Chaozhou: both are naturally conformable without measuring and cutting.
>
> 好作奇語自是文章病，但當以理為主。理得而辭順，文章自然出羣拔萃。觀杜子美到夔州後詩，韓退之自潮州還朝後文章，皆不煩繩削而自合矣.[23]

In contradistinction from Lü Dafang's tripartite periodization, Huang Tingjian suggests a before-and-after model. With Lü Dafang, the evolution of a writer's 'verbal vitality' is parallel to, and in synchrony with, the biological progression from youth to prime to old age. In contrast, Huang Tingjian pinpoints a certain significant biographical event as the artistic watershed, such as Du Fu's move to Kuizhou in 766 and Han Yu's return from his exile in Chaozhou in 820. Thus, to the temporality of late style is added a spatial dimension, which is informed by political frustration or personal hardship experienced by the author.

The three adjectives Lü Dafang uses to characterize the styles of great writers at different stages of their lives are all positive in meaning, though one may infer that 'rigorous' represents the highest level of accomplishments. In Huang Tingjian's before-and-after model, there is a more explicit and abrupt hierarchy: the 'after' presents as a quantum leap instead of a gradual progress from the 'before.' The relationship between the two phases becomes contrastive rather than transitional or evolutional. There is also a difference between Lü Dafang and Huang Tingjian in their characterizations of late style. Whereas the former foregrounds technical rigor, the latter stresses a kind of free conformity to law. The phrase 'naturally conformable without any measuring and cutting' invokes Confucius's famous account of the culmination of his life-long moral cultivation at the age of seventy, when he was able to 'follow my heart's desire without transgressing any rules' 隨心所慾不越矩.[24]

It may be noted here that Du Fu himself was one of the first to broach the subject of lateness, especially with regard to the sixth-century poet Yu Xin庾信 (513–581). In 554, Yu Xin was sent by the Southern Liang court on a mission to the Western Wei in Chang'an, where he was detained for the rest of his life and where he witnessed the devastating fall of the Liang. Du Fu repeatedly linked the pathos of Yu Xin's late works to the travails of his life:

appended to Chen Xiang, *Guling xiansheng wenji* 古靈先生文集 (Collected Works of Master Guling), in *Beijing tushuguan guji zhenben congkan* 北京圖書館古籍珍本叢刊 (A Collection of Treasured Editions of Ancient Works from Beijing Library), 120 vols. (Beijing: Shumu wenxian chubanshe, 1988), vol. 87, 242.

23 Huang Tingjian, 'Yu Wang Guanfu shu yi' 與王觀復書·一 (Letter to Wang Guanfu, No. 1), *QSW*, 104:2281.297.

24 See *Lunyu* 論語 (Analects), 2.4. In literary critical discourse, 'naturally conformable without measuring and cutting' had been used by Han Yu in praise of Fan Zongshi's 樊宗師 (726?–824) writings. See Han Yu, 'Nanyang Fan Shaoshu muzhi ming' 南陽樊紹述墓誌銘 (Tombstone Epitaph for Fan Shaoshu of Nanyang), *Han Yu wenji jiaozhu* 韓愈文集校註, eds. Liu Zhenlun 劉真倫 and Qiu Zhen 丘珍 (Beijing: Zhonghua shuju, 2010), 25.2575.

庾信平生最蕭瑟	Yu Xin's life the most dismal time of all:
暮年詩賦動江關	in twilight years his poems and *fu* stirred the River Pass.[25]

Corresponding to Yu Xin's life experience are the stylistic as well as conceptual changes in his late poetry:

庾信文章老更成	Yu Xin's literary works grew more perfect in old age,
凌雲健筆意縱橫	a sturdy brush topping the clouds, his concepts moved at will.[26]

Du Fu would also occasionally dole out generous praise to friends and acquaintances in reference to their late poetry, even though there is no other indication that they were poets of high caliber. In a poem addressed to Xue Qu 薛璩, for example, after marveling at the physical stamina of his friend in old age, he turns to his poetic prowess:

賦詩賓客間	You compose poems among the guests,
揮灑動八垠	what you dash off stirs the world's eight bounds.
乃知蓋代手	Now I understand that a master of a whole age,
才力老益神	grows more divine in force of talent as he ages.[27]

More important, Du Fu suggested that something drastic happened to his own poetry as he aged. A poem composed in 761 (when he was fifty-two) mentions his poetic transformation:

為人性僻耽佳句	As a person I'm one-sided by nature, addicted to lovely lines,
語不驚人死不休	if my words don't startle others, in death I will not quit.
老去詩篇渾漫興	As I get older my poems are pretty much easy-going,
春來花鳥莫深愁	when spring comes the birds and flowers don't upset me very deeply.[28]

Despite the daunting ambiguity in the fourth line, it is abundantly clear that Du Fu's relentless pursuit of striking phraseology gave way to a spontaneous expression of impromptu moods in his late years. This might contradict Lü Dafang's description of Du Fu's late works as 'rigorous.' However, Du Fu also tells us that 'In my later phase I gradually get more precise with poetry's rules' 晚節漸於詩律細.[29]

One may conclude that formality and licentiousness are two sides of the same coin in Du Fu's late works, but it is important to keep in mind that his remarks were occasion-bound and isolated from contemporary critical discourse. He did not intend to articulate a coherent theory; nor should we try to construct one based on his desultory remarks. The discourse on lateness in the Song, on the other hand, exhibits a clear tendency towards theoretical schematization and dogmatic rigidity. This tendency became most pronounced in Fang Hui's 方回 (1227–1307) comments on Du Fu in his anthology entitled *Yingkui lüsui* 瀛奎律髓 (Finest Poems in Regulated

25 'Singing My Feelings On Traces of the Past' 詠懷古跡 (1 of 5), *The Poetry of Du Fu*, 4:361–362.
26 'Six Quatrains Done Playfully' 戲為六絕句 (No. 1), *The Poetry of Du Fu*, 3:112–113.
27 'Sent to Director Xue Ju (3)' 寄薛三郎中璩, *The Poetry of Du Fu*, 5:96–97.
28 'By the River I Came on Waters that Looked Like a Sea: A Short Account' 江上值水如海勢聊短述, *The Poetry of Du Fu*, 3:16–17.
29 'Getting Rid of the Blues, Playfully Shown to Vice-Director Lu [19]' 遣悶戲呈路十九曹長, *The Poetry of Du Fu*, 5:68–69.

Verse by Tang-Song Luminaries), completed in 1283, just four years after the official fall of the Southern Song. The following is a sample:

> Generally speaking, in Du Fu's collection, the poems he wrote in Chengdu (759–765) are superior to those he wrote in Guan-Fu (750–759); those he wrote in Kuizhou (766–768) are superior to those in Chengdu; those in Hunan (768–770) are superior to those in Kuizhou. [His poetry is like a plant growing] taller with each joint: the older it grows, the more it peels and sheds off.
>
> 大抵老杜集，成都時詩勝似關輔時，夔州時詩勝似成都時，而湖南時詩又勝似夔州時，一節高一節，愈老愈剥落也.[30]

In Fang Hui's description, the organicist metaphor is turned upside down. The evolution of Du Fu's poetry becomes an essentializing process that naturally and continuously gets rid of flowery verbiage. The idea of the-older-the-better is pushed to its logical—and some might argue absurd—conclusion. The minute distinction of Du Fu's various mini-phases in accordance with his movement from one place to another may seem a bit fussy and mechanical, but, as Ji Yun 紀昀 (1724–1805) has pointed out, it is an offshoot of Huang Tingjian's two-stage model, which Ji Yun dismisses as 'heroic bluffing' 英雄欺人.[31]

The elevation of Du Fu's late poetry was part of an emerging and expanding discourse on lateness at the turn of the twelfth century.[32] The paradigm for evaluating Du Fu's works soon found its way into critical and self-critical reflections on major Song poets. Following Huang Tingjian, it became a common practice to identify an equivalent Kuizhou moment for other poets. Liu Kezhuang 劉克莊 (1187–1269) likened Wang Anshi's retirement to Jiangning in 1077 to Du Fu's move to Kuizhou.[33] Hu Zi 胡仔 (fl. 1147–1167) drew a comparison between Su Shi's 蘇軾 (1037–1101) poems written after his exile to Lingnan in 1094 and Du Fu's post-Kuizhou poems: both epitomize what is meant by 'rigorous in old age.'[34] Zhu Bian 朱弁 (1085–1144) identified two transformations in Su Shi. The first occurred with Su Shi's exile to Huangzhou in 1080, after which his poetry became superb so that only Huang Tingjian could occasionally rise to his height. His banishment to the Hainan Island seventeen years later marked the inception of his last phase, when even Huang Tingjian would not be able to come anywhere close to him.[35] In regard to Du Fu's post-Kuizhou poems and Su Shi's post-Lingnan writings, Fang Hui cited the comment of an anonymous scholar that 'Their mellow brush outshines their juvenile writings. Even their works in their middle years cannot match those of the late years' 老筆愈勝少作，而中年亦未若晚年也.[36]

30 Fang Hui, *Yingkui lüsui huiping* 瀛奎律髓匯評 (Finest Poems in Regulated Verse by Tang-Song Luminaries, with Collected Comments), ed. Li Qingjia 李慶甲 (Shanghai: Shanghai guji chubanshe, 1986), 10.325. For a similar comment by Fang Hui, see ibid., 20.780.

31 Fang Hui himself referred to Huang Tingjian's statement on Du Fu as authoritative. See his 'Cheng Doushan yingao xu' 程斗山吟稿序 (Preface to Cheng Doushan's Poetry Manuscript), *QSW*, 7:212.90.

32 For a perceptive discussion of this topic, see Zhou Yukai 周裕鍇, *Songdai shixue tonglun* 宋代詩學通論 (A General Discussion of Song Dynasty Poetics) (Shanghai: Shanghai guji chubanshe, 2007), 346–354.

33 Liu Kezhuang, 'Liu Qifu shi xu' 劉圻父詩序 (Preface to Liu Qifu's Poetry), *QSW*, 329:7565.79.

34 Hu Zi, *Tiaoxi yuyin conghua, houji* 後集 (Later Collection), 30.226.

35 Zhu Bian, *Fengyuetang shihua* 風月堂詩話 (Remarks on Poetry from Hall of Wind and Moon), ed. Chen Xin 陳新 (Beijing: Zhonghua shuju, 1988), 1.106.

36 *Yingkui lüsui huiping*, 16.615.

Huang Tingjian's case is particularly noteworthy in that he not only devised the two-stage model for evaluating Du Fu's poetry but also applied it in delineating his own poetic pilgrimage. Just two years before his death, in an 1103 letter to a nephew, he took stock of his writing career and pointed to his exile to Qianzhou in 1094 as the defining moment: Before that he did not understand the art of writing so that what he wrote in those days all appeared laughable to him now. It was only after he arrived at Qianzhou that he began to understand how to write.[37] Commentators would follow his lead in singling out his post-Qianzhou works as his poetic best. Thus, Cai Tao 蔡絛 (11th–12th cent.) noted 'a transformation of style in his poetry' 詩變前體 after he returned from Qianzhou.[38] An anonymous twelfth century commentator described how his creative power was set free after he arrived at Qianzhou so that his 'art of verse' 句法 became so matchlessly superb among Song poets.[39]

The before-and-after model was also applied in prose criticism. A fine example is the following remark by Han Biao 韓淲 (1159–1224) on five prose masters of the Song (all of whom would later be placed in the pantheon of the Eight Grand Masters of Prose of the Tang and Song 唐宋八大家):

> After 'Account of the Pavilion of the Drunken Old Man,' Duke Ouyang's writings mellowed tremendously. After 'Account of Snow Hall,' Zizhan's [Su Shi] writings got rid of all mannerisms as found in pieces written for civil service exams. After Jiefu [Wang Anshi] was relieved of his duty as Grand Councilor and returned to Banshan, the power of his brush became extremely lofty and archaic. After Zeng Zigu [Zeng Gong 曾鞏 (1019–1083)] met Duke Ouyang, his writings became far superior to others. In the writings of Su the Senior [Su Xun 蘇洵 (1009–1066)], there is not a single piece that shows traces of refinement through axing and chiseling. That is because he had burned all his juvenile writings.
>
> 歐陽公自《醉翁亭》後，文字極老。蘇子瞻自《雪堂》後，文字殊無制科氣象。介甫之罷相歸半山也，筆力極高古矣。如曾子固見歐陽公後，自是迴出於諸人之上。老蘇文字，篇篇無斧鑿痕，蓋少作皆已焚之矣。[40]

Ouyang Xiu 歐陽修 (1007–1072) (Duke Ouyang) composed his account in 1046, at the age of thirty-nine. Su Shi was forty-five when wrote his piece in 1082. Wang Anshi's retirement took place in 1077, when he was fifty-six. Zeng Gong was only twenty-two years old when he first met Ouyang Xiu in 1041. The wide range of these dates indicates that a transformation can occur at any time in a writer's life, from youth (Zeng Gong) to prime time (Ouyang Xiu and Su Shi) to old age (Wang Anshi). At the same time, the reference to Su Xun suggests that Han Biao is still framing the transformation in terms of the dichotomy between youth and old age. The grip of this dichotomy on the critical mind was such that the transformative moment was often automatically or unwittingly conceived of as the gateway to old age, as seen in an entry on

37 Huang Tingjian, 'Da Hong Jufu shu' 答洪駒父書·二 (Letter in Reply to Hong Jufu, No. 2), *QSW*, 104:2281.301.

38 Cai Tao, *Xiqing shihua* 西清詩話 (Remarks on Poetry from the Clean Western Chamber), in *Xijianben Songren shihua si zhong* 稀見本宋人詩話四種 (Rare Editions of Four Song Remarks on Poetry), ed. Zhang Bowei 張伯韋 (Nanjing: Jiangsu guji chubanshe, 2002), 208.

39 Hu Zi, *Tiaoxi yuyin conghua, houji*, 32.245, quoting anon., 'Yuzhang xiansheng zhuan zan' 豫章先生傳贊 (Encomium to Master Yuzhang).

40 Han Biao, *Jianquan riji* 澗泉日記 (Jianquan's Diaries), in *Quan Song biji* 全宋筆記 (Complete Brush Notes of Song Dynasty), ed. Zhu Yian 朱易安 (Zhengzhou: Daxiang chubanshe, 2003–), sixth series, vol. 9, 3.123.

'Perfection of Poetry in Old Age' 老而工詩in Sun Yi's 孫奕 (b. 1126) *Lüzhai shier bian* 履齋示兒編 (*A Collection by Lüzhai for His Children*):

> A guest said, 'The artistry of a poet does not need to be evaluated in accordance with whether he was in his youthful prime or mellow old age.' I said, 'That is not true. In the poems by Drunken Old Man [Ouyang Xiu] after he arrived at Yiling and those by Oldster of Fu [Huang Tingjian] after he arrived at Qiannan, allegory and symbolism became all the more brilliant, and the use of allusions became all the more intricate. Their short pieces were elegantly grand; their long pieces were heroically archaic. The poems by Shaoling [Du Fu] after he arrived at Kuizhou and those by Changli [Han Yu] after his exile to Chaoyang became all the more brilliant. Otherwise, why would Shaoyou [Qin Guan 秦觀 (1049–1110)] say that 'A Poem on the Sagely Virtues of the Yuanhe Era' is an inferior piece in Han [Yu's] writings and that it seems to have come from a different hand when compared with 'Stele Inscription on the Pacification of Huaixi'? Simply because that was a juvenile piece.
>
> 客有曰：「詩人之工于詩，初不必以少壯老成較優劣。」　余日：殆不然也，醉翁在夷陵後詩，涪翁到黔南後詩，比興益明，用事益精，短章雅而偉，大篇豪而古。如少陵到夔州後詩，昌黎在潮陽後詩，愈見光焰也。不然，少遊何以謂《元和聖德詩》于韓文為下，與《淮西碑》如出兩手，蓋其少作也。」[41]

By the second half of the twelfth century, Kuizhou, Chaozhou, and Qianzhou had been well established as the watershed moments in the respective writing careers of Du Fu, Han Yu, and Huang Tingjian. Sun Yi, however, created something of an oddity in designating the exile to Yiling as Ouyang Xiu's moment. The exile took place in 1036, when Ouyang Xiu was barely thirty. The oddness becomes even more conspicuous when we consider that Han Yu's poem was composed in 807, when he was already thirty-nine. In other words, whereas Han Yu, who was by no means developmentally impaired, was still in his 'juvenile' stage at thirty-nine, Ouyang Xiu already reached his 'mellow old age' when he was only twenty-nine.

Repudiating Juvenility and Elevating Lateness

Qin Guan's discrimination between Han Yu's poem and prose piece highlights the contrast between juvenility and mellowness.[42] In such a contrast, the repudiation of juvenility emerges as a necessary and inevitable correlative to the elevation of lateness. This correlation can be best illustrated through an examination of the critical assessment of Wang Anshi's poetry. The following remark by Ye Mengde 葉夢得 (1077–1148) serves as a convenient point of departure:

> When young, Duke Wang [Anshi] of Jing prided himself on his aspirations. His poetic language simply followed his disposition without any reserve. . . . Later, after he became Administrative Assistant in the Herds Office, he borrowed from Song Cidao [Song Minqiu 宋敏求 (1019–79)] all his collections of Tang poets. He read widely among those and was very selective in choosing his models. Not until his late years, however, did he comprehend the subtlety of profound indirection and unforcednesss. One can conclude thereby that,

41 Sun Yi, *Lüzhai shier bian* (Beijing: Zhonghua shuju, 2014), 10.166.

42 Qin Guan's remark was first recorded in Chen Shidao 陳師道 (1053–1101), *Houshan shihua* 後山詩話 (Houshan's Remarks on Poetry), in *Lidai shihua* 歷代詩話 (Remarks on Poetry from Various Dynasties), comp. He Wenhuan 何文煥 (1732–1809) (Beijing: Zhonghua shuju, 1981), 309.

> although a man's writing abilities are predetermined, they still depend on his age. Even his lordship could not forcefully hasten to his destination before his time came.
>
> 王荆公少以意氣自許，故詩語惟其所向，不復更爲涵蓄……後爲羣牧判官，從宋次道盡假唐人詩集，博觀而約取，晚年始盡深婉不迫之趣．乃知文字雖工拙有定限，然亦必視初壯，雖此公，方其未至時，亦不能力强而遽至也。[43]

In this influential narrative, Wang Anshi's poetry went through a transformation during his tenure in the Herds Office (from 1054 to 1057) when he was in his thirties, though Ye Mengde did not offer any details of this middle style other than suggesting that it was indebted to Tang models. Wang Anshi's retirement from office in 1077 ushered in his last phase, when he attained the ideal of 'the subtlety of profound indirection and unforcednesss.' Schematically, this tripartite periodization harkens back to Lü Dafang, but there are significant differences. With Lü Dafang, biography and biology are connected but remain distinct: the former generates the content; the latter plays a more crucial role in shaping the style. In Ye Mengde's account, age and experience combine to determine a poet's style at a given stage of his development. For Lü Dafang, biography consists largely of the subject's socio-political experiences. Ye Mengde, on the other hand, foregrounds the role of literary cultivation—it was Wang Anshi's intense engagement with Tang poetry that brought about his first transformation. Perhaps the most significant difference between the two lies in their characterization of the changing styles of their subjects: The adjectives that Lü Dafang uses to describe Du Fu's three stages are all positive in meaning. Ye Mengde's elevation of Wang Anshi's late works, on the other hand, necessitates the repudiation of his early ones.

From the second decade to the middle of the twelfth century, the world of letters was dominated by the mutually admiring coterie of Ye Mengde, Zhai Ruwen 翟汝文 (1076–1141), Wang Zao 汪藻 (1079–1154), and Sun Di 孫覿 (1081–1169).[44] It is all but certain that Sun Di was echoing Ye Mengde in a letter to Zeng Zao 曾慥 (d. 1155) around 1150,[45] though there is a minor concern when exactly Wang Anshi's second phase began. In that letter, Sun Di denunciates the poor taste of his contemporaries, which was manifested, among other things, in the popularity of some of Wang Anshi's early poems, even though they are not really 'fine verses' 佳句. In Sun Di's estimate, Wang Anshi's poems did not achieve 'extreme perfection' 盡善 until after he became a Drafter of Imperial Decrees 知制誥 (in 1061). It was after he retired from office that he 'fashioned a subtle and exquisite style' 造精微. What he wrote thereafter 'is a world apart from his juvenile writings' 比少作如天淵相絶也.[46] It should be noted here that the simultaneous praise of Wang Anshi's late works and dispraise of his early ones was not confined to the Ye Mengde circle. It was rather an emerging critical consensus. Li Gongyan李公彦 (1079–1133), for example, also remarked that the poems that Wang

43 Ye Mengde, *Shilin shihua jiaozhu* 石林詩話校注 (Shilin's Remarks on Poetry, Collated and Annotated), ed. Lu Mingxin 逯銘昕 (Beijing: Renmin wenxue chubanshe, 2011), 2.93.

44 See Gong Mingzhi 龔明之 (1091–1182). *Zhong Wu jiwen* 中吳紀聞 (Record of What Was Heard from the Central Wu), in *Quan Song biji*, 3rd series, vol. 7, 7.261.

45 Sun Di's letter was written upon receipt of a copy of Zeng Zao's *Song baijia shixuan* 宋百家詩選 (Selections from a Hundred Song Poets), composed sometime between 1148 and 1151. For the date of Zeng Zao's anthology, see Lu Sanqiang 陆三强, 'Zeng Zao san kao' 曾慥三考 (A Study of Three Aspects of Zeng Zao's Life), in *Gudai wenxian yanjiu jilin* 古代文献研究集林 (Collections of Studies of Ancient Materials), 2nd series, ed. by Huang Yongnian 黄永年 (Xi'an: Shaanxi shifan daxue chubanshe, 1992), 214–215.

46 Sun Di, 'Yu Zeng Ruibo shu' 與曾端伯書 (Letter to Zeng Ruibo), *QSW*, 159:3429.54.

Anshi wrote in retirement are 'exquisitely profound and marvelously subtle, to which his juvenile writings cannot compare' 精深華妙，非少作之比.[47]

The criticism of Wang Anshi's early poems was a pushback against their contemporary popularity. It also reveals the anxiety, particularly acute among the Ye Mengde circle, that the excellence of Wang Anshi's late works might be obscured. Exacerbating this anxiety was the lopsided circulation of his poetry in the early decades of the Southern Song, as documented in Wang Zao's colophon to *Banshan bieji* 半山別集 (A Separate Collection of Banshan). The collection was originally compiled and printed in the early 1100s by Zhu Ting 祝廷 (*jinshi 1079*).[48] It contained some 100 poems and ten memorials 表 and letters 啟 composed by Wang Anshi in his late years. It was then reprinted by Zhou Yanzhi 周彥直, a former student of Wang Anshi. Xu Fu 徐俯 (1075–1141) reported that, when Huang Tingjian read the poems in the collection, he 'would clap and applaud at every single line' 句句擊節. However, for twenty years after the Song court moved to the south, Wang Zao was unable to obtain a copy. When he read a collection of Wang Anshi's works entitled *Linchuan qian hou ji* 臨川前後集 (A Two-Part Collection of Linchuan), he recognized therein dozens of poems from Zhu Ting's collection. (None of the memorials and letters in that anthology had survived.) He picked out these poems to form a separate collection in part to rectify a sad situation; namely that people were only talking about Wang Anshi's 'juvenile writings' 少作, whereas his 'mellow brush'老筆 was no longer appreciated.[49]

To understand the specific qualities in Wang Anshi's early works that were targeted by critics, it is helpful here to take a closer look at one of the examples cited by Sun Di, i.e., the second couplet in 'Together with My Younger Brother I Chanted on the Bamboos at the Pavilion of These Gentlemen in the Temple of Glorious Treasures' 與舍弟華藏院此君亭詠竹 (*QSS*, 559.6643):

一逕森然四座凉 A single path into the lushness, cool all around;
殘陰餘韵興何長 In slanting shadow and lingering rustle, our mood knows no bound.
人憐直節生來瘦 People cherish them for their straight joints: naturally thin;
自許高材老更剛 They take pride in their tall timber: sturdier in old age.
曾與蒿藜同雨露 Having shared rain and dew with wormwood and goosefoot,
終隨松柏到冰霜 They will accompany pine and cypress in ice and frost.
煩君惜取根株在 Prithee do not take them away with their roots and trunk;
欲乞伶倫學鳳凰 I wish to ask Ling Lun to mimic the phoenix's call.[50]

47 See Hu Zi, *Tiaoxi yuyin conghua qianji, 33.222, quoting Mansou shihua* 漫叟詩話 (Mansou's Remarks on Poetry). For the attribution of *Mansou shihua to Li Gongyan, see Guo Shaoyu* 郭紹虞, *Song shihua kao* 宋詩話考 (A Study of Remarks on Poetry from the Song) (Beijing: Zhonghua shuju, 1979), 147–149.
48 The precise date of Zhu Ting's anthology is not known. As it was compiled while he served in the Bureau of Education 學事司 in Huainan, it can be dated between 1103 (when the government formally established the Bureau of Education at the circuit level) and Huang Tingjian's death in 1105. For the establishment of the Bureau of Education at the circuit level, see Too 脫脫 (1313–55), *Songshi* 宋史 (History of the Song Dynasty) (Beijing: Zhonghua shuju, 1985), 167.3971; Gong Yanming 龔延明, *Songshi zhiguanzhi buzheng* 宋史職官志補正 (Additions and Emendations of the Treatises on State Offices in the History of the Song Dynasty) (Hangzhou: Zhejiang guji chubanshe, 1991), 454–456.
49 Wang Zao, 'Ba Banshan shi' 跋半山詩 (Colophon to Banshan's Poems), *QSW*, 157:3384.239.
50 Ling Lun, founder of music, fashioned bamboo flutes. He was inspired to devise the twelve pitch-standards when he heard the phoenix's call at the foot of Kunlun Mountains. See *Lüshi chunqiu jiaoshi* 呂氏春秋校釋 (The Annals of Mr. Lü), ed. Chen Qiyou 陳奇猷 (Shanghai: Xuelin chubanshe, 1984), 5.284–285.

The poem was probably written around 1038, when Wang Anshi was still in his teens.[51] The younger brother was Wang Anguo 王安國 (1028–1074), who, ten years old at the time, also composed a (now-lost) poem.[52] The name of the pavilion derives from an anecdote about Wang Huizhi 王徽之 (338–386), who, when asked why he would bother to have bamboos planted in another man's vacant house where he was lodging temporarily, said that he could not be without 'these gentlemen' even for a single day.[53] As perennial evergreen plants, bamboos emblemize moral perseverance in Chinese culture; with their hollow insides, they also suggest mental receptiveness. Here, Wang Anshi projects an explicit self-image by turning the bamboos into a symbol of virtue and talent. What Sun Di found aesthetically undesirable was precisely the explicitness and even aggressiveness in Wang Anshi's use of easily recognizable symbolism.

There is anecdotal evidence that in his late years Wang Anshi himself felt uneasy about the couplet in spite of, or rather because of, its popularity.[54] He would show displeasure when people praised it to his face. Once, while sitting in the pavilion with Wang Anguo, he gazed at the inscription of the couplet on a plaque and regretted that his youthful composition, in wide circulation now, could no longer be changed. Wang Anguo compared his brother's regret to that of Yang Xiong 揚雄 (53 BCE–18 CE), who famously dismissed his juvenile rhapsodies.[55] For Li Bi 李壁 (1159–1222), the problem of the couplet lies in its style: 'Its language does not possess the charm of suggestiveness; that is why it is regrettable' 語無含蓄風韻，故當悔之.[56] Ji Yun's critique is similar: A 'poem with a physical thing as its topic' 詠物詩 would be shallow if there were no 'allegorical import' 比興. If such import were too obvious, on the other hand, it would be equally shallow.[57]

If the anecdote about Wang Anshi's regret is true, then he may be considered the first major Song poet to repudiate his own juvenile works. Such self-repudiation would soon become common. In his late years, Huang Tingjian, as we have mentioned, dismissed his writings before the Shaosheng era as laughable. In fact, already in his middle years, he burned two-thirds of the more than one thousand poems he had written in his younger days.[58] The dramatic gesture of

51 Wang Anshi was living with his father Wang Yi 王益 (d. 1039), who was serving as the Controller-general 通判 of Jiangning. See Wang Jinguang 王晉光, *Wang Anshi shi xinian chugao* 王安石詩系年初稿 (A Draft on the Dates of Wang Anshi's Poems) (Manila: Deyang gongsi, 1986), 1–2. Li Deshen 李德身, on the other hand, suggests that the poem was written in 1066, when Wang Anshi was staying in Jiangning before he went to Kaifeng to assume office at court; see his *Wang Anshi shiwen xinian* 王安石詩文系年 (Dates of Wang Anshi's Poetry and Prose) (Xi'an: Shaanxi renmin jiaoyu chubanshe, 1987), 174.

52 Wang Anguo was child prodigy. At the age of eleven *sui*, he already established a literary reputation with his dozens of compositions in the various genres. See Wang Anshi, 'Wang Pingfu muzhi' 王平甫墓誌 (Tombstone Epitaph for Wang Pingfu), *QSW, 65:1413.131; Zeng Gong* 曾鞏 (1019–1083), 'Wang Pingfu wenji xu' 王平甫文集序 (Preface to the Literary Collection of Wang Pingfu), *QSW, 58:1523.2.*

53 Liu Yiqing 劉義慶 (403–44), *Shishuo xinyu jianshu* 世說新語箋疏 (A New Account of Tales of the World, with Annotations and Sub-Annotations), ed. Yu Jiaxi 余嘉錫, revised edition (Shanghai: Shanghai guji chubanshe, 1993), 23.759.

54 A reflection of this popularity is Liu Ban's 劉攽 (1023–1089) poem 'The Pavilion of These Gentlemen in the Temple of Precious Treasures in Jianye' 建業華藏院此君亭 (*QSS*, 610.7230), which, as its short preface states, was inspired by Wang Anshi's poem.

55 Hu Zi, *Tiaoxi yuyin conghua qianji*, 34.229, quoting Zeng Zao, *Gaozhai shihua* 高齋詩話 (Gaozhai's Remarks on Poetry). For Yang Xiong's dismissal of the rhapsodies he composed in his youth, see *Fagan yishu* 法言義疏, annotated by Wang Rongbao 王榮寶 (Beijing: Zhonghua shuju, 1987), 'Wu zi' 吾子 (Our Masters), 45. Yong Xiong's rejection of his early writings derived from his doubt about their moral validity whereas Wang Anshi's regret was prompted by aesthetic consideration.

56 Li Bi, *Wang Jing Wengong shi jianzhu* 王荊文公詩箋注 (The Poetry of Duke Wang of Jing, with Annotations and Sub-annotations), ed. by Gao Keqin 高克勤 (Shanghai: Shanghai guji chubanshe, 2010), 33.830.

57 Ji Yun's comment is included in Fang Hui, *Yingkui lüsui huiping*, 35.1425.

58 Ye Mengde, *Bishu luhua* 避暑錄話 (Sayings Recorded While I Sought Relief from Summer Heat), in *Quan Song biji*, 2nd series, vol. 10, 10.272–273. In a letter of 1060 to Ouyang Xiu, Su Xun 蘇洵 (1009–66) mentions that he, after carefully

burning one's earlier poems often signified the determination to take a new path. After Chen Shidao 陳師道 (1053–1101) met Huang Tingjian, he burned all the thousands of poems he had written before—he would turn a new leaf by following Huang Tingjian.[59] In 1162, when Yang Wanli 楊萬里 (1127–1206) was in his mid-thirties, he burned all his youthful poems (totaling more than a thousand) because they were mostly works in the style of the Jiangxi School of poetry, which he came to reject after he found new models in Chen Shidao, Wang Anshi, and late Tang poets.[60]

When referring to Ye Mengde's critique of Wang Anshi's early works, modern scholars often cite such ancient-style poems as 'Disarmament' 省兵, 'Monopolizers' 兼併, and 'Opening up Relief Granaries' 發廩. It must be pointed out that ancient-style poems dealing with social and political issues form a subgenre in which direct expression is accepted and even expected. The explicitness in such poems is not the concern of Ye Mengde. His focus is on Wang Anshi's modern-style poetry, especially the construction of the parallel couplet. It is in modern-style poetry that Wang Anshi is said to have reached perfection late in life:

> In his late years, Duke Wang of Jing was particularly precise and rigorous with poetry's rules. There is not even a hair's breadth of laxity in his choice of words and use of characters. Meaning merges with language; language is deployed according to meaning. The fusion is so natural that you do not see any place where things are forcefully pulled or fixed together.
>
> 王荊公晚年詩律尤精嚴，造語用字，間不容髮。然意與言會，言隨意遣，渾然天成，殆不見有牽率排比處。[61]

One of the three examples cited by Ye Mengde is the second couplet of 'South Bank' 南浦 (*QSS*, 564.6687):

南浦東岡二月時 From South Bank to East Mount, in this second month,
物華撩我有新詩 The splendor of things spurs me into writing a new poem.
含風鴨綠鱗鱗起 Embracing the wind, duck-green ripples, ever so blithe;[62]
弄日鵝黃裊裊垂 Frolicking with the sun, goose-yellow dangles, always pliant.

---

perusing the writings of ancients, burned dozens of the writings he had composed. See his 'Shang Ouyang Neihan diyi shu' 上歐陽內翰第一書 (The First Letter to Palace Writer Ouyang), *QSW, 43:919.27.* What Su Xun burned were prose pieces rather than poems.

59 Chen Shidao, 'Da Qin Gou shu' 答秦覯書 (A Letter in Reply to Chen Gou), *QSW*, 123:2664.285.

60 Yang Wanli, 'Chengzhai Jiang Hu ji xu' 誠齋江湖集序 (Preface to Chengzhai's Collection of Jiangxi and Hunan), *QSW*, 238:5321.218. It should be noted here that the act of burning one's early or earlier poems should not always be viewed in the context of the discourse on lateness. Throughout his life, Chen Shidao would burn whatever he had written if there was anything in it that he found dissatisfactory so that only about ten percent of his writings had survived. See Xie Kejia 謝克家 (d. 1134), 'Houshan jushi ji xu' 後山居士集序 (Preface to the Collected Works of Houshan Jushi), *QSW*, 145:3133.319. In Xie Kejia's account, Chen Shidao's manuscript-burning displayed his diligence in weeding out what is not worth preserving in his literary collections. Similarly, He Zhu 賀鑄 (1052–1125), would review his old writings every three years and burn those that he considered to be unworthy. See He Zhu, 'Qinghu Yilao shiji xu' 慶湖遺老詩集 (Preface to the Poetry Collection of the Old Man of Qinghu), *QSW, 124:2673.48.*

61 Ye Mengde, *Shilin shihua jiaozhu*, 1.12.

62 All three extant editions of Wang Anshi's poetry have 粼粼 (clear) in place of 鱗鱗. I follow Ye Mengde in adopting 鱗鱗, which also appears in the other two *shihua* texts from the late Northern and early Southern Song, i.e., Wei Tai, *Lin Han yinju shihua jiaozhu*, 2.99, and Huihong, *Lengzhai yehua* 冷齋夜話 (Night Talk in Cold Studio), ed. Chen Xin 陳新 (Beijing: Zhonghua shuju, 1988), 4.37.

The parallel structure in the couplet is so obvious that one may puzzle at Ye Mengde's observation that 'When you first read it, you won't realize there is parallelism' 讀之初不覺有對偶. What he intends to stress are the hidden intricacies in Wang Anshi's 'choice of words and use of characters.' One may immediately recognize the standard parallelism between 'duck' and 'goose' in the category of waterfowls and that between 'green' and 'yellow' as color words, but there are less apparent forms of verbal manipulation. 'Duck-green' (the dark green color of a duck's head feather) works as a metonym (*daiyu* 代語) for the verdant water. Similarly, 'goose-yellow,' the light yellow color of a gosling's down, is a substitute for budding willows. Such employment of metonym is highly valued in Song criticism. 'In alluding to things and refining your verse lines,' according to Huihong 惠洪 (1071–1128), 'the subtlety lies in mentioning the manifestations of something without mentioning its name' 用事琢句，妙在言其用，不言其名耳. The cleverness of Wang Anshi's couplet lies precisely in describing the manifestations of water and willows without mentioning their names.[63]

The juxtaposition of the two descriptive binomes *lingling* 鱗鱗 (blithe) and *niaoniao* 裊裊 (pliant) illustrates the subtlety of Wang Anshi's verbal maneuver down to the level of radicals. Here, the fish radical 魚 and the bird radical 鳥 constitute a sub-parallelism, known as *cedui* 側對 (lateral parallelism), as fish and birds are routinely paired to form a parallel in Chinese poetry and prose. With the parallel construction in the couplet informed with all the intricacies at the levels of words, characters, and radicals, one can appreciate Wang Anshi's declaration that it 'almost surpasses the Creator of things' 幾淩轢造物.[64]

Ye Mengde saw Wang Anshi's technical perfection in both eight-line regulated verse poems and quatrains. In the course of the twelfth century, it was Wang Anshi's quatrains that increasingly came to be regarded as the cream of his late works. A statement attributed to Huang Tingjian proved to be the most seminal: 'The small poems that Duke Jing [Wang Anshi] composed in his twilight years are gracefully elegant and superbly refined; they completely transcend the run-of-the-mill and the vulgar. Every time I intone them with relish, I feel sweet dew coming out between my tongue and teeth' 荊公暮年作小詩，雅麗精絶，脱去流俗，每諷味之，便覺沆瀣生牙頰間.[65] There is strong evidence that Huang Tingjian's original statement, which has *xiaoyu* 小語 (small pieces) instead of *xiaoshi* 小詩 (small poems), was actually about Wang Anshi's short letters on Chan Buddhism.[66] Nonetheless, the statement became the critical cornerstone in establishing the supremacy of Wang Anshi's quatrains. Hu Zi clearly understood the term 'small poems' as referring to quatrains when he, after [mis]quoting Huang Tingjian, cited as examples six pentasyllabic quatrains, which he described as 'truly capable of making one chant and sigh again and again' 真可使人一唱而三歎也.[67]

---

63 Huihong, *Lengzhai yehua*, 4.37.

64 Wang Zhifang 王直方 (1069–1109). *Wang Zhifang shihua* 王直方詩話 (Wang Zhifang's Remarks on Poetry), in Guo Shaoyu 郭紹虞, *Song shihua jiyi* 宋詩話輯佚 (A Collection of Excerpts from the Lost Remarks on Poetry in the Song) (Beijing: Zhonghua shuju, 1980), 61. In Li Bi, *Wang Jing Wengong shi jianzhu, 41.1046,* 造物 (Creator of things) reads as *chunwu* 春物 (spring objects). When Wei Tai 魏泰 (b. 1041?), during a visit in the spring of 1083, asked him about his latest poetic compositions, he proudly responded by chanting the poem. See Wei Tai, *Lin Han yinju shihua jiaozhu*, 2.99.

65 Hu Zi, *Tiaoxi yuyin conghua qianji*, 35.234, quoting Huihong, *Lengzhai yehua*.

66 See Huang Tingjian, 'Ba Wang Jinggong Chan jian' 跋王荊公禪簡 (Colophon to Duke Wang of Jing's Short Letters on Chan Buddhism), *QSW*, 106:2310.109: 'These small pieces from his twilight years are gracefully elegant and superbly refined. They transcend the popular and the vulgar and cannot be approached from the perspective of common sense' 暮年小語，雅麗精絶，脱去流俗，不可以常理待之也.

67 Hu Zi, *Tiaoxi yuyin conghua qianji*, 35.234.

Puwen 普聞 (fl. 1130) also singled out Wang Anshi's quatrains as his crowning achievement:

> Dongpo [Su Shi] excels in ancient-style poems with his heroic abandon and magnificent grandeur; Luzhi [Huang Tingjian] excels in the regulated verse poems with his mellow power and transcendent sublimity; Duke Jing excels in quatrain poems with his easy elegance and pristine sparseness. Each is a master of his domain. However, Duke Jing's poems are profound and deliberate. That is what people value nowadays.
>
> 東坡長於古韻，豪逸大度；魯直長於律詩，老健超邁；荊公長於絕句，閒雅清臒。其各一家也。然則荊公之詩覃深精思，是亦今時之所尚者。[68]

Puwen described Wang Anshi's poetic excellence as fragmented at three levels: formally, it is confined to the quatrain, stylistically to 'easy elegance and pure sparseness,' and temporally to his late years. The example cited by Puwen is 'There Was a Light Rain as at Longan Where I Bid Adieu to Hefu, So I Wrote This to Send to Mrs. Wu, Née Wang' 送和甫至龍安微雨因寄吳氏女子 (*QSS* 567.6711):

荒煙涼雨助人悲 Desolate mist and chilly rain added to our sadness;
染濕衣衿不自知 Unbeknownst to us, our robes dampened.
除卻春風沙際綠 But for the spring wind that greened the sandy banks,
一如送汝過江時 It felt exactly the same as when I sent you across the river.

Wang Anshi wrote the poem in 1082, on the occasion of seeing off his brother Wang Anli 王安禮 (1024–1095) (whose polite name was Hefu) on his way to the capital for an official appointment.[69] Longan was the name of a ferry, just west of Jiangning, where Wang Anshi had lived in retirement. Mrs. Wu was Wang Anshi's eldest daughter, married to Wu Anchi 吳安持. She had not seen her father for seven years and suffered from acute homesickness, as expressed in a quatrain that she sent to her father in the fall of 1081.[70] Wang Anshi's response expressed understanding of her sadness but tried to convince her that separation is a fact of life. He urged her to read Buddhist sutras so as to transcend her secular attachment.[71]

What distinguishes the present quatrain from Wang Anshi's previous poems to his daughter is its intense sentimentality without any pretension to philosophical profundity. Puwen's comment is that it 'brushes aside the appearance of heroic abandon and sweeps away the mannerism of mellow power. Its meaning and resonance are profound and far-reaching. It achieves both pristine sparseness and elegant brilliance' 拂去豪逸之氣，屏蕩老健之節，其意韻幽遠，清臒雅麗為得也。[72] The quatrain may be too lachrymose to be representative of Wang Anshi's late works, but Puwen's approach is noteworthy. Unlike Ye Mengde and Sun Di who vertically compared Wang Anshi's late works with his early ones, Puwen engaged in horizontal comparisons with Su Shi and Huang Tingjian in terms of form and style. For him,

68 Puwen, *Shilun* 詩論 (A Discussion of Poetry), in *Song shihua quanbian* 宋詩話全編 (Complete Remarks on Poetry of the Song Dynasty), 10 vols., ed. Wu Wenzhi 吳文治 (Nanjing: Jiangsu guji chubanshe, 1998), 2:1426–1427.
69 For other poems by Wang Anshi written on the occasion, see 'Song Hefu zhi Longan mugui' 送和甫至龍安暮歸 (Returning in Twilight, after Seeing Off Hefu at Longan), *QSS*, 567.6706.
70 See 'Ji fu' 寄父 (Sent to My Father), *QSS*, 893.10445.
71 See 'Ci Wushi nü yun' 次吳氏女韻 (Harmonizing with the Poem by Mrs. Wang), *QSS*, 568.6713. In 'Ji Wushi nüzi' 寄吳氏女子 (Sent to Mrs. Wu, Née Wang), *QSS*, 565.6696, Wang Anshi drew on Daoist teaching to convey the same idea.
72 Puwen, *Shilun*, 2:1427.

style is very much tied with, if not entirely determined by, verse form: a quatrain would simply be too short to exhibit the kind of 'heroic abandon and magnificent grandeur' found in Su Shi's ancient-style poems or the 'mellow power and transcendent sublimity' exhibited in Huang Tingjian's regulated verse poems.

The quatrain cited by Puwen is heptasyllabic rather than pentasyllabic, though there is otherwise no evidence that he prefers Wang Anshi's heptasyllabic quatrains to his pentasyllabic ones (for which Hu Zi apparently had a predilection). Zhang Bangji 張邦基 (fl. 1131) put Wang Anshi on equal footing with Tang poets explicitly on the strength of his heptasyllabic quatrains, which he described as 'limpid and graceful' 清婉.[73] For Yang Wanli, Wang Anshi's heptasyllabic quatrains served as a model for emulation. By his own account, he first studied the masters of the Jiangxi school, then the pentasyllabic regulated verse poems of Chen Shidao, then the heptasyllabic quatrains of Wang Anshi, and finally the quatrains of Tang poets. Later, he experienced a kind of epiphany and realized the importance of intuitive and instantaneous comprehension of the principles of poetry instead of a step-by-step approach through emulating earlier models.[74] Nonetheless, he maintained a visceral enjoyment of Wang Anshi's quatrains.[75]

Occasional dissentions notwithstanding, the status of Wang Anshi's quatrains became firmly established in the course of the twelfth century.[76] One critic went so far as to declare him one of the two best quatrain writers in history, the other being the Tang poet Du Mu 杜牧 (803–53).[77] It was in such a context that Yan Yu 嚴羽 (1191–1241), in a chapter titled 'Shi ti' 詩體 (Poetic Modes) in his *Canglang shihua* 滄浪詩話 (Canglang's Remarks on Poetry), employed the term *Wang Jinggong ti* 王荊公體 (the style of Duke Wang of Jing). By *ti* Yan Yu referred to a variety of things, including poetic genre, verse form, period style, group style, and individual style. He did not provide any definition of *Wang Jinggong ti*, but he did append a note to the term to indicate that Wang Anshi's style is best embodied in his quatrains: 'The Duke's quatrains are the most outstanding. Where he is at his best, he surpasses Su Shi, Huang Tingjian, and Chen Shidao but there remains a barrier between him and Tang poets.'[78] Implicitly but clearly, Wang Anshi's late style is defined in terms of his achievement in a particular poetic form.

Concluding Remarks

I would like to conclude by reflecting briefly on four dimensions of the idea of lateness as it developed in Song China: 1) the historicity of the idea in light of its antecedent and concomitant counter-discourse of ageism, 2) its limited applicability as a critical concept, 3) the defining qualities in the late works of great writers, and 4) hypotheses about the genesis and formation of late style. I will make brief references to the discourse on lateness in the West in the hope

73 Zhang Bangji, *Mozhuang manlu* 墨莊漫錄 (Mozhuang's Desultory Jottings) (Beijing: Zhonghua shuju, 2002), 6.180.
74 Yang Wanli, 'Chengzhai Jingxi ji xu' 誠齋荊溪集序 (Preface to Chengzhai's Collection of Jing Stream), *QSW*, 238:5321.219.
75 See Yang Wanli's 1190 poem 'Du shi' 讀詩 (Reading Poems), *QSS*, 2305.26489.
76 For a prominent critique of Wang Anshi's quatrains, see Ye Shi 葉適 (1150–1223). *Xixue jiyan xumu* 習學記言序目 (Study Notes on Classics and Histories) (Beijing: Zhonghua shuju, 1977), 47.707.
77 Zeng Jili 曾季貍 (b. 1118?), *Tingzhai shihua* 艇齋詩話 (Tingzhai's Remarks on Poetry), in *Lidai shihua xubian* 歷代詩話續編 (A Sequel Collection of Remarks on Poetry from Various Dynasties) ed. Ding Fubao 丁福保 (Beijing: Zhonghua shuju, 1983), 299.
78 Yan Yu, *Canglang shihua jiaojian* 滄浪詩話校箋 (Canglang's Remarks on Poetry, Collated and Annotated), ed. Zhang Jian 張健. Shanghai: Shanghai guji chubanshe, 2012), 2.240.

that a cross-cultural comparative perspective may prompt us to take a fresh look at what would otherwise remain familiar or even taken for granted.

The Song discourse on lateness was preceded by an ageist narrative that associates the last phase of a writer with decline in talent. Early illustrations can be found in anecdotes about three prominent fifth-century poets. In his late years, Bao Zhao 鮑照 (414–466) adopted 'lowly language and faulty sentences' 鄙言累句 in his writing so that he would not undermine Emperor Xiaowu 孝武 (r. 854–864), who declared himself matchless in the art of writing. People at the time thought that 'his talent had run out' 才盡.[79] Ren Fang 任昉 (460–508) was revered as a prose writer, just as Shen Yue 沈約 (441–513) was for his poetry. He could not, however, live with the thought that his poetic skills were inferior to Shen Yue's. Late in his life, he turned his passion to poetic composition in the hope of outshining Shen Yue but his heavily allusive language lacked smooth flow. 'As a result, there was talk that his talent had run out' 於是有才盡之談矣.[80] Neither Bao Zhao nor Ren Fang truly experienced a decline in literary capability in old age: the former performed a political act of self-preservation; the latter was led astray by his literary hubris. Nonetheless, the default conclusions reached by their contemporaries betray an unspoken assumption that a writer's talent is exhaustible and that the exhaustion tends to occur late in life. Such assumption is best embodied in the legend of Jiang Yan 江淹 (444–505). At the age of fifty-six, Jiang Yan had a life-changing dream. In that dream, a man asked him to return a five-colored writing brush that had been left in his care for years, and he complied with the request. Thereafter, he lost his ability to write fine poems. Consequently, people all thought 'Jiang Yan's talent had run out' 江淹才盡.[81] He has since become a symbol of the decline of talent in old age.

The discourse on lateness was not only preceded but also countered by the ageist discourse in the Song period. Zhu Xi 朱熹 (1130–1200), for example, roundly rejected the premise that the best of a great writer is always achieved late in life. He debunked in this regard the myth surrounding Du Fu's Kuizhou poems, which he found to be inferior to many of the poet's earlier works.[82] He recognized that a momentous life experience may prompt a drastic change in writing style, such as Han Yu's exile to Chaozhou and Su Shi's to Hainan, but he would not accept that all great writers get better as they age.[83] In his view, a writer's 'vital style' 氣格 is formed before he reaches thirty; any difference thereafter is simply a matter of refinement.[84] He maintained that 'when a man gets old, his vitality declines, so does the quality of his writing' 人老氣衰，文亦衰.[85] He compared the writing of an old man to calligraphy done with a bald brush: both have lost their 'sharpness' 鋒銳.[86] Han Biao

79 This happened after Bao Zhao received an appointment at court in 456. See Shen Yue 沈約 (441–513), *Song shu* 宋書 (History of Liu-Song Dynasty) (Beijing: Zhonghua shuju, 1974), 51.1480.

80 Li Yanshou 李延壽 (active 7th cent). *Nan shi* 南史 (History of the Southern Dynasties) (Beijing: Zhonghua shuju, 1978), 59.1455.

81 Zhong Rong 鐘嶸 (468?–518?), *Shipin zhu* 詩品注 (Grading of Poets, with Annotations), ed. Chen Yanjie 陳延傑 (Beijing: Renmin wenxue chubanshe, 1961), 49–50.

82 Echoing Zhu Xi, Ji Yun would observe that in general Du Fu's middle poems are his best whereas his late works tend to appear too 'dejected' 頹唐. See Fang Hui, *Yingkui lüsui huiping*, 10.325.

83 Zhu Xi, *Zhuzi yulei* 朱子語類 (Thematic Discourses of Master Zhu), comp. Li Jingde 黎靖德 (13th cent.) (Beijing: Zhonghua shuju, 1986), 140.33.

84 Zhu Xi, *Zhuzi yulei, 139.3301.*

85 Zhu Xi, *Zhuzi yulei*, 139.3311. The example Zhu Xi mentioned was Ouyang Xiu. Ouyang Xiu himself once lamented the decline of his talent in old age. See his 'Ti Qingzhou shanzhai' 題青州山齋 (Written on My Mountain Cottage in Qingzhou), *QSW, 34:718.93.*

86 Zhu Xi, *Zhuzi yulei*, 139.3302

took a more nuanced position by noting that some writers peak at middle age and deteriorate thereafter while others continue to improve in old age. Of the five Song prose masters whose transformative moments he identified, two (Ouyang Xiu and Zeng Gong) belong in the first category and three (Su Shi, Wang Anshi, and Su Xun) in the second.[87]

In the West, the discursive rise of lateness has a much more recent origin, but it was similarly preceded by a counter-discourse. Up to the mid-nineteenth century, it had been generally assumed that creativity reached its highest point in maturity and declined thereafter. The various 'Ages of Man' schemas invariably posited old age as a period of deterioration. In the second half of the nineteenth century, the idea began to gain currency that the late works of great writers and artists exhibited certain aesthetic qualities distinct from and superior to those found in their earlier works. By the beginning of the twentieth century, 'late style' had been established 'as a given, as a hallmark of genius and as a transhistorical phenomenon.'[88]

Just as the Chinese idea of lateness germinated in the canonization of Du Fu, its Western counterpart first developed in the re-evaluation of the last works of two canonical masters: Beethoven and Goethe. The canonicity of lateness restricts its critical applicability to a relatively small number of geniuses. It is nonetheless a capacious concept that is applied broadly in musical, literary, and fine arts criticism in the West.[89] Great artists, composers, and writers from different historical periods and cultural backgrounds such as Titian, Shakespeare, Rembrandt, Beethoven, Goethe, Ibsen, Picasso, and Stravinsky are described as 'sharing certain qualities in their late works.'[90] In the Chinese tradition, on the other hand, the idea of lateness is severely limited at three levels. First of all, it is employed almost exclusively in literary criticism.[91] This narrowness has to do with the lack of a canon of composers or fine artists whose extant works can be evaluated in connection with their biographies. Secondly, within literary criticism, the idea is confined to the canonical genres of poetry and prose; fiction and drama are excluded because their status remained low throughout the premodern period. Finally, within each canonical genre, there is a further restriction: late style is embodied almost exclusively in the *shi* 詩 form of poetry and ancient-style prose.

In the West, a great artist's final phase of production is 'characterized in one of two modes: either as serene, synthetic, and consummatory, or as irascible, discordant, and recalcitrant (and sometimes in a curious, contradictory combination of the two).'[92] It is 'the aesthetic values of the critic that in the end determine what is deemed positive or negative in the last works of an artist. Therefore, if you are a Georg Simmel, you appreciate wholeness, coherence, synthesis; older age becomes a time of reappraisal, summary, consolidation. But if you are a modernist (and Marxist-inflected) Adorno or a Said, you treasure fragmentation,

87 Han Biao, *Jianquan riji*, 3.124.

88 Gordon McMullen and Sam Smiles, 'Introduction: Late Style and Its Discontents,' in *Late Style and Its Discontents: Essays in Art, Literature, and Music* (ed. Gordon McMullen and Sam Smiles; Oxford: Oxford University Press, 2016), 3. See also Michael Hutcheon and Linda Hutcheon, 'Late Style(s): The Ageism of the Singular,' *Occasion: Interdisciplinary Studies in the Humanities* 4 (2012), 4: 'the history of late style as a critical concept, while first articulated in the classical and Renaissance periods, actually emerged in its current forms early in the twentieth century, but as conceptualized through the work of a series of earlier German idealist and romantic thinkers.'

89 Gordon McMullen observes that the idea of lateness made its impact felt first not in literary criticism, but in musicology. See his *Shakespeare and the Idea of Late Writing: Authorship in the Proximity of Death* (Cambridge: Cambridge University Press, 2007), 2,

90 McMullen and Smiles, 'Introduction,' 3.

91 In *Songdai shixue tonglun*, 353–354, Zhou Yukai mentioned a few examples in Song painting criticism. These isolated examples do not seem sufficient to form the basis on which to construe the idea of lateness in painting criticism.

92 McMullan and Smiles, 'Introduction,' 3–4.

dissonance, lack (or impossibility) of reconciliation.'[93] As one scholar puts it, lateness is 'a construct, ideological, rhetorical and heuristic, a function not of life or of art but of the practice of reading or appreciating certain texts within a set of predetermined parameters.'[94]

In general, Chinese critics would have agreed with Simmel more than Adorno or Said. Du Fu, as we have seen, described his late poems in two opposing terms: 'precise with poetry rules' and 'easy-going.' These two terms morphed respectively into Lü Dafang's 'rigorous in old age' and Huang Tingjian's 'naturally conformable without measuring and cutting.' Later commentators would pick and choose one of the two to describe late style. Hu Zi used 'rigorous in old age' to characterize Su Shi's post-Lingnan poems as well as Du Fu's post-Kuizhou poems. Yuan Haowen 元好問 (1190–1257), on the other hand, observed that Du Fu's poems after Kuizhou, Bai Juyi's after Xiangshan, and Su Shi's after Hainan are all 'naturally conformable without measuring and cutting.'[95] Wang Anshi's late works have been described in a wider variety of terms: 'subtlety of profound indirection and unforcednesss' and 'precise and rigorous with poetry's rules' (Ye Mengde); 'exquisitely profound and marvelously subtle' (Li Gongyan); 'completely transcend the run-of-the-mill and the vulgar' (Huang Tingjian); 'truly capable of making one chant and sigh again and again' (Hu Zi); 'profound and deliberate' and 'pristine sparseness and elegant brilliance' (Puwen); 'limpid and graceful' (Zhang Bangji). The qualities described by these terms may be different but not irreconcilable.

The most significant difference between China and the West lies in the ways in which lateness is situated. There are two models in the Chinese tradition. In the tripartite scheme devised by Lü Dafang, lateness is linked with biological and biographical progression into old age. A variation is Ye Mengde's emphasis on literary cultivation and cumulative learning. In Huang Tingjian's two-stage theory, a momentous event is marked as the entry into lateness, such as Du Fu's move to Kuizhou, Zeng Gong's and Chen Shidao's meeting with a master, and the retirement from office in the case of Bai Juyi and Wang Anshi. The event that is most often identified as the catalyst is political exile, such as Han Yu's to Chaozhou, Ouyang Xiu's to Yiling, Su Shi's to Huangzhou, Huizhou, and the Hainan Island, and Huang Tingjian's to Qianzhou. Such a conception of exilic experience dovetails with the idea of poetic perfection through adversity 窮而後工 prevalent in the Song period.[96] The thematic thread in all the theories is that the transformation into lateness is a function of something that happened. In contrast to this retrospective approach, the interpretative framework in the West tends to be prospective: lateness is believed to be 'conditioned by the artist's particular physical and psychological circumstances' with the proximity to death.[97] In other words, lateness in the Chinese tradition is described as an outcome of a past occurrence; in the West, it is conceived of as a response to (or anticipation of) something that is yet to come.

---

93 Michael and Linda Hutcheon, 'Late Style(s), 5.

94 McMullen, *Shakespeare and the Idea of Late Writing*, 5.

95 Yuan Haowen, 'Taoran ji shi xu' 陶然集詩序 (Preface to Taoran Collection), in *Quan Yuan wen*, 1:20.316–317.

96 The phase was first used by Ouyang Xiu in 'Mei Shengyu shiji xu' 梅聖俞詩集序 (Preface to Mei Shengyu's Poetry Collection), *QSW, 134:716.52. The idea itself had a much older history. For a critical review, see Gong Bendong* 鞏本棟, ''Qiong er hou gong' de lishi kaocha' '詩窮而後工'的歷史考察, Z*hongshan daxue xuebao, 2008 (4):19–27. For dissenting views idea in the Song period, see Guan Qin* 管琴, 'Songdai 'qiong er hou gong' lun zhi yishuo kao 宋代'窮而後工'論之異說考 (An Investigation of the Dissenting Views of the Theory of 'Perfection Through Adversity' in the Song Dynasty), *Wenyi yanjiu*, 2016 (12):73–82. For a cross-cultural comparative discussion, see Qian Zhongshu 錢鍾書, 'Shi ke yi yuan' 詩可以怨 (Poetry as Expression of Resentment), in his *Qi zhui ji* 七綴集 (A Collection of Seven Essays), 2nd edition (Shanghai: Shanghai guji chubanshe, 1994), pp. 119–138.

97 McMullan and Smiles, 'Introduction,' p. 4.

The Evolution of Terracotta from the First to the Second Half of the II Millennium BC

Laura Battini

It's a great pleasure for me to contribute to the 100th Anniversary of this noble American Oriental Society with an article that attempts to give new insights for the understanding of Kassite production.[1] Studying terracottas is already not an easy topic. Certainly, most of the articles or books on the subject concern the Old Babylonian period,[2] but none deals with the continuity of this production into the Kassite period and even further.[3]

First of all, studies on ancient objects are still influenced by the modern canonical distinction between art on one hand and craft-production on the other.[4] Little attention is given even today to terracotta production. Of course, the distinction between 'art' and 'craft' has no resonance in the Ancient Near East, where not one 'artist' has left his signature.[5] There, tradition and conformity to tradition were basic elements in production (artistic as well as craft),[6] in contrast to art today where 'innovation' and 'originality' have been the artistic aims since Greek times. In particular the studies on Kassite production in clay are linked more to aesthetical considerations than to the significance of this production inside the society where it developed. For instance, the comparison of two clay heads, an Old Babylonian god's head

1 I am very indebted to Richard Beal and JoAnn Scurlock for having read my communication in the conference organized in February 2017 at St. Mary's College in Notre Dame Indiana. But I am very indebted to them for the kindness and the refreshing discussions they always have with colleagues of all backgrounds, ranks and ages.

2 Ruth M. Opificius, *Das altbabylonische Terrakottarelief* (Untersuchungen zur Assyriologie und Vorderasiatischen Archäologie 2; Berlin: W. de Gruyter, 1962); Nadja Cholidis, *Möbel in Ton, Untersuchungen zur archäologischen und religionsgeschichtlichen Bedeutung der Terrakottamodelle von Tischen, Stühlen und Betten aus dem Alten Orient* (Altertumskunde des Vorderen Orients 1; Münster: Ugarit Verlag, 1992); Elizabeth Stone, 'Chariots of the Gods in Old Babylonian Mesopotamia (2000-1600 BC),' *Cambridge Archaeological Journal* 3/1 (1993): 83-10; Elise Auerbach, 'Terra Cotta Plaques from the Diyala and their Archaeological and Cultural Contexts,' (PhD Dissertation; University of Chicago, 1994); Julia Assante, 'Style and Replication in 'Old Babylonian' Terracotta Plaques: Strategies for Entrapping the Power of Images', in *Ex Mesopotamia et Syria Lux. Festschrift für Manfried Dietrich* (ed. Oswald Lorenz, et al.; Alter Orient und Altes Testament 281; Münster: Ugarit Verlag, 2002), 1–25; Laura Battini, 'La déesse aux oies : une représentation de la fertilité?' *Revue d'Assyriologie* 100 (2006): 57-70.

3 Apart from Marie-Thérèse Barrelet, *Figurines et reliefs en terre cuite de la Mésopotamie antique* (Bibliothèque Archéologique et Historique de l'Institut Français d'Archéologie d'Istanboul 85; Paris: P. Geuthner, 1968), 95-109. But it is often difficult to identify Kassite production (cf. Evelyn Klengel-Brandt and Nadja Cholidis, *Die Terrakotten von Babylon im Vorderasiatischen Museum Berlin: Die anthropomorphen Figuren* (Wissenschaftliche Veroffentlichung der Deutsche Orient-Gesellschaft 115; Saarwellingen: Saarländische Druckerei und Verlag, 2006), 21-22; Glenn Schwartz, 'The Value of the Vestigial. From Middle to Late Bronze Age in Ebla and Western Syria,' in *Ebla and Beyond* (ed. Paolo Matthiae, F. Pinnock and M. D'Andrea; Wiesbaden: Harrassowitz, 2018), 439-474. Cf. also Agnès Spycket, 'Un naos à divinité bovine,' in *Beiträge zur Altorientalischen Archäologie und Altertumskunde. Festschrift für Barthel Hrouda zum 65. Geburtstag* (ed. Peter Calmeyer *et al.*; Wiesbaden: Harrassowitz, 1994), 269–72.

4 Carole Talon-Hugon, *L'esthétique.* (Paris: Presses Universitaires de France, 2008), Isabelle de Maison Rouge, *L'art contemporain* (Paris, Le Cavalier bleu, 2002), Georges Roque, *Majeur ou mineur? Les hiérarchies en art* (Nîmes: J. Chambon, 2000). For a criticism of the use of this modern aesthetic in the Ancient Near East see Laura Battini, 'Popular Art and Official Art: a Possible and Useful Classification in Mesopotamian Iconography?' *Res Antiquitatis* 4 (2013): 1-20. And on the exchanges between the two kinds of production see Laura Battini, 'Iconography Exchanges between ' Popular ' and ' Elite ' 'Minor Arts', in *Proceedings of the 8th International Congress on the Archaeology of the Ancient Near East;* (ed. P. Bieliński *et al.;* Wiesbaden: O. Harrassowitz, 2014), 453-466.

5 Irene Winter, 'Aesthetics and Mesopotamian Art,' in *Civilizations of the Ancient Near East* (ed. Jack Sasson; New York: Charles Scribner's Sons, 1995), 2569-2580; Winfied Orthmann, 'Aspects of the Interpretation of Ancient Near Eastern Art as Visual Communication,' in *Proceedings of the 4th International Congress on the Archaeology of the Ancient Near East* (ed. Hartmut Kühne; Wiesbaden: Harrossowitz, 2008), 243-254.

6 Winter, 'Aesthetics', Orthmann, 'Aspects', Battini, 'Iconography Exchanges'.

Figure 1. Old Babylonian god's head and a Kassite man's head

and a Kassite man's head (fig 1), not only demonstrates that an esthetic judgment is not justified – both heads are so naturalistic and finely executed--but it also demonstrates a real continuum between the first and the second half of the 2nd millennium B.C.[7] In fact, as in the Old Babylonian period, similarly in the Kassite period both official and 'popular' arts used two styles, one linear, which is the most widespread, and one more naturalistic.

Secondly, most of articles dealing with clay figurines and plaques concern the Old Babylonian period because it is the moment when they are by far the most numerous. Indeed, the Kassite period is one of the most obscure in Mesopotamian history; at the present, the documentation is still comparatively scanty in material culture, in official artwork and in texts.[8] Little attention has been given to Kassite production and in particular to the relationship, between the first and the second half of the 2nd millennium BC. In fact, judgments on Kassite production often depend on reductive comparisons with the Old Babylonian period.[9] Imbued with classical culture, most archaeologists judge Kassite production as rough, dull, poor... These judgments constitute a

7 The period of the passage between Middle Bronze and Late Bronze is still problematic (see most recently Schwartz, 'Value of the Vestigial') especially in Mesopotamia, where few excavations concern the Late Bronze Age: cf. Klengel-Brandt and Cholidis, *Terrakotten*, 21-22.

8 Cf. Agnès Spycket, 'Kassite and Middle Elamite Sculpture' in *Later Mesopotamia and Iran. Tribes and Empires 1600-539 BC.* Seminar in Memory of Vladimir G. Lukonin (John Curtis ed.; London: British Museum Press, 1995), 25-32. Two hundred royal inscriptions – most of them short– sixty *kudurru*-s and approximately 12,000 letters and economic texts are not a large amount of knowledge, especially when one thinks that less than 10% has been published. There is no doubt that archaeological excavations will uncover more documents.

9 Since Elisabeth Van Buren, *Clay Figurines of Babylonia and Assyria* (Yale Oriental Series Researches 16; New Haven: Yale University Press, 1930), nos. 126, 161, 185-186, 466, 512-3, 523, 588, 954-956, 958, 1026-1027, 1135-1138, 1181, 1264-1265)

serious prejudice to the study of the Kassite terracottas and prevent a calm reconsideration of this production.[10]

Thirdly, even the Old Babylonian terracottas are still not well understood, as I demonstrate in the course of this paper.[11] In such a case, how is it possible to judge Kassite production by comparison with an Old Babylonian production that is not well understood ? The risks of misunderstanding are enormous.

In the following analysis I use an historical method in order to link the clay production to the society and to the human needs for which it was created. This can, on one hand, avoid value judgments, in particular stylistic and aesthetic ones, and on the other give new inputs and perspectives of analysis. Terracottas are an excellent source for understanding the life of ancient people: they show religious feelings, troubles, worries of every day life, and also they can shed light on popular toys and the lucky charms. We don't ask today for toys or for charms to have an esthetic appearance. In the same way, the most useful way to study clay artifacts is to relate them to the human needs for which they were created.

I. The Kassite Terracotas

Kassite terracotta are known from at least three sites: Dur Kurigalzu (Aqar Quf), Nippur and Babylon.[12] The figurines found at Nippur and Babylon are very few in number. By way of contrast, Aqar Quf seems to have produced more than one hundred examples. However, their exact number is uncertain and only a few have been published and photographed.[13] Preliminary reports are very short, not accurate and selective. Only 43 figurines were published. They represent only human and animal subjects, no divinities, no models, no plaques. The humans are primarily heads, rather than full bodies. Perhaps this is an impression based on the restricted number of figurines published, and/or on the artistic desire of the photographer, who has favored human physiognomy. At Nippur, on the other hand, full-body human figurines have been found.

The first element that is quite surprising, after having a look at all the published material, is the difference between figurines from the capital (Aqar Quf) and those from other sites (fig. 2). In the capital, the head of the human figurines is depicted in more detail, with a particular attention to the eyes. Colors highlight the care taken in the depiction of human heads. Sometimes traces of colors are found on the eyes and hair, but it is possible that certain figurines were covered by colors everywhere. Animal figurines are also formed with more attention to detail; also there, the colors stress the naturalistic. Eyes are highlighted by colors (dark) in both human and animal figurines. Other details are rendered by incisions. The most common animal is a

10 With the exception of Rita Dolce, 'Per una identificazione della plastica cassita: le origini' in Francesca Baffi-Guardata and Rita Dolce, *Archeologia della Mesopotamia. L'età cassita e medio-assira* (Archeologica 88; Rome: G. Bretschneider editore, 1990).

11 And as I showed in a communication 'Terracottas from Old Babylonian Ur,' delivered at the 65e Rencontre Assyriologique Internationale (2016) in Philadelphia (in press).

12 Mohammed Ali Mustafa, 'Kassite Figurines. A New Group Discovered Near 'Aqar Qûf,' *Sumer* 3 (1947): 19-22, Taha Baqir, 'Iraq Government Excavations at 'Aqar Quf., Second Interim Report, 1943-1944,' *Iraq Suppl.* 2 (1945): 1-15 and idem, 'Iraq Government Excavations at 'Aqar Quf., Third Interim Report, 1944-1945,' *Iraq* 8 (1946): 73-93; Dolce, 'plastica cassita'; Klengel-Brandt and Cholidis, *Terrakotten*; and Richard L. Zettler, *Nippur III. Kassite Buildings in Area WC-1* (Oriental Institute Publications 111; Chicago: The Oriental Institute, 1993).

13 The preliminary field reports do not give an exact number of the discoveries. Also, when photographed, the quality is so bad that it is sometimes difficult to have an exact idea of the figurine (especially in Baqir, 'Second Interim Report' and idem, 'Third Interim Report').

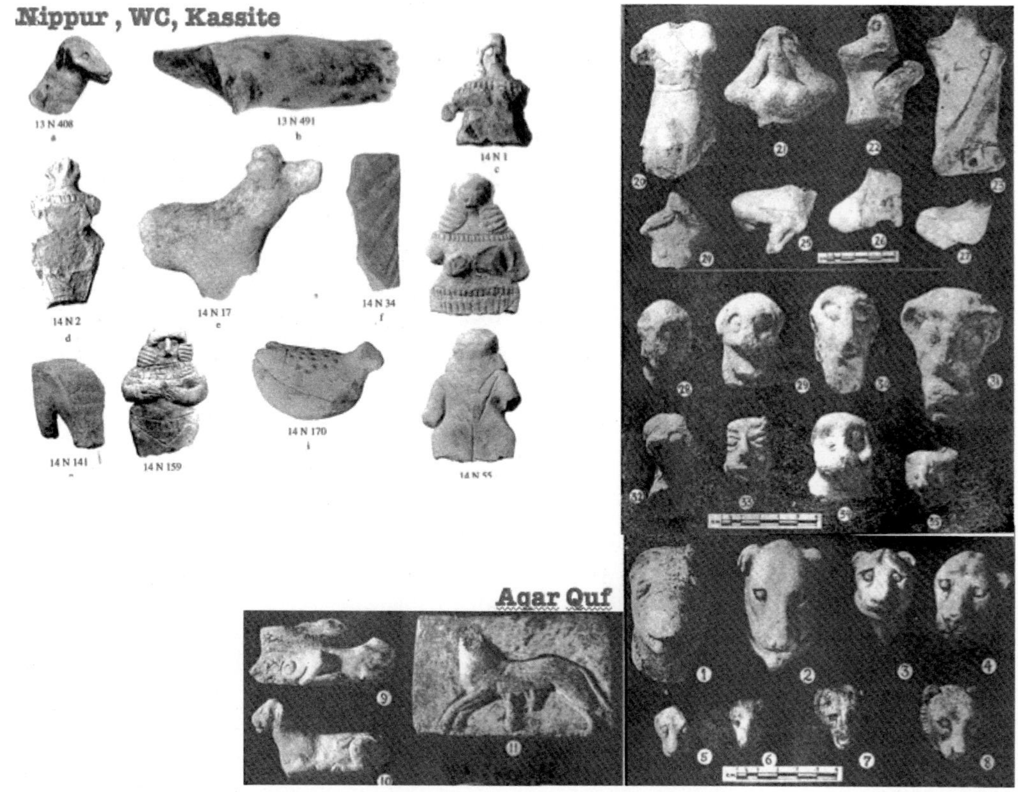

Figure 2. Figurines from Aqar Quf and from other sites

feline at Aqar Quf; on the contrary, sheep are more frequent at Nippur and Babylon. Animals in Aqar Quf are more naturalistic than those from Nippur.

The second quite remarkable element is the proportion between female and male figurines: males are more numerous than females, at least in Aqar Quf.[14] This is in contrast to what is known of clay figurines of the 3rd millennium and also of the beginning of the 2nd millennium B.C. A third fascinating characteristic of Kassite clay-figurine production is the presence of inscriptions on the body of the figurine, both animal and human. It would be very interesting to know the content of these inscriptions, but unfortunately they have not yet been published. By comparison with the 1st millennium inscribed terracottas, it might be possible to suggest a protective function for the Kassite inscribed figurines, which would suggest a continuity from the second half of the 2nd millennium B.C. to the 1st millennium B.C.

II. Comparisons between Kassite and the Old Babylonian production

Old Babylonian clay-figurine production is much better known than Kassite. If only three Kassite sites have brought to light terracottas, the number is five times greater in the Old Babylonian period, when at least 15 sites have attested clay-figurine production. The difference is also quantitative (table I): each Old Babylonian site has produced an important group of terracottas.

14 In Kassite Nippur female figurines remain more numerous (Zettler, *Nippur* III, pl. 56).

Girsu (Tello), Uruk and Ur are the richest ones (between 300 and 450 terracottas), but also at Nippur, Kish, Ešnunna (Tell Asmar), Babylon, Isin and Nērebtum (Tell Ischali) a large number of clay figurines were found (between more than one hundred and 250). In total, 2513 terracottas are known from the first half of the 2nd millennium, while in the second half of the same millennium, the total is little more than one hundred.

Even the find-spots are far more varied in the Old Babylonian than in the Kassite period.[15] Old Babylonian terracottas (table II) were found in houses, temples, palaces, graves and workshops. Houses, palaces and workshops are also attested in the Kassite period. No other contexts are known from this period, probably because of the far fewer number of sites yielding figurines. There is no doubt that future excavations will increase not only the number of figurines but also the different types of find-spots.[16]

The material and technique remain unchanged in the two periods (table III): clay is the simplest material, used for everything in southern Mesopotamia. Economically it is important for manufacturing objects like terracottas, which are destined for everyone, that the raw material not be expensive. Most of the Old Babylonian terracottas are made serially with moulds, but there are also hand-made figurines. In absence of data from the reports of excavations, and looking at the plates published, the impression is that Kassite figurines are not made in moulds – and, in fact. no mould has ever been found for this period – but by hand. Regarding color, it is difficult to judge because this topic is only taken into account in recent years and no analysis is available for figurines, even those of the Old Babylonian period. But in both periods color was used; the question that remains to be solved is if it was used on all or only on a few examples. Another difference with the Old Babylonian period is the reduction of typologies. In the first half of the 2nd millennium, production in clay included figurines, plaques, chariots and little furniture -like beds, tables, and chairs. In Kassite Mesopotamia, on the contrary, clay objects are limited to figurines, human and animal, but no plaques and no models are for the moment attested. Also, the variety of female figurines and even male figurines, contrasted with the variety of subjects known from Old Babylonian period, is drastically reduced (table IV). There is no representation of a god, no representation of the king, no representation of demoniac or protective spirits. What was still produced is the master of the animals and the naked frontal female – a subject very much debated and not yet understood in all periods. There are also new subjects not attested in the Old Babylonian period manufactures, as for example the female who hides her eyes with her hands.

III. Difficulties of comprehension of clay artifacts

Since we still do not understand the Old Babylonian terracottas, how could we possibly understand those of the Kassite period, which are so badly published? Archaeologists know in general only the most common themes of Old Babylonian terracottas. And even some very common themes, such as that of the naked frontal woman, are really difficult to understand.

15 The context of discovery doesn't have to mean that the terracottas were always where they were found. Rather, all this means is that they were at a certain moment in that location. Indeed, apart from the context of graves, the only one for which it's sure that the figurines were deposited voluntarily there, the figurines are really very easy to transport, even if archaeologists are unconcerned by this subject. But it's an important one, because it suggests new functions for terracottas. I return to this also in the conclusions.

16 For example, we know of the presence of terracottas under urban streets only for the 1st millennium B.C. (Babylon: Robert Koldewey, *Das wieder erstehende Babylon* (Leipzig: J.C. Hinrichs, 1913) and for the 3rd millennium. (Mari: Jean Margueron, *Mari: métropole de l'Euphrate, au IIIe et au début du IIe millénaire av JC.* (Paris: Picard, 2004) and Béatrice Muller, 'Mari, chantier F (2000): la 'maquette architecturale' D,' *Akh Purattim* 3 (2015): 261-276). So their absence during the 2nd millennium is probably fortuitous and new excavations will fill this void.

Spycket 1994 fig.1
Isin 1967

Barrelet 1968 n.132
From Tello

Figure 3. Old Babylonian terracottas representing rare themes

In the same way, if erotic plaques are read very easily, their functions escape us in large part. But even more difficult is the case of Old Babylonian terracottas which represent rare themes (fig.3). For example, the terracotta showing a divinity having a bull head is attested three times in Mesopotamia,[17] twice from Isin, and once from the antiquities market. It is possible that it is a representation of the goddess Ištar or one of her hypostases, but it is not known why the terracottas of this subject appear only for a very short time and (almost?) all from one city. Another example is the god with bull ears, considered sometimes as a dying divinity.[18] But in reality for this figure there is not yet a convincing theory, nor explication. And it is strange that so few archaeologists are concerned with these not well understood terracottas. I'm convinced that these ones are the most intriguing and that they better allow us to improve our understanding of ancient society.

The functions of the terracottas of any period are also unknown. For example, one of the most striking aspects of these terracottas is their portability: small in size and not very heavy, they could be easily transported.[19] This means a great mobility — unexpected until very recently—

17 Laura Battini,'La terre cuite IB 1967 ou peut-on lire les images coroplastiques à travers les textes?' *Iraq* 71 (2009): 125-138.

18 Barrelet, *Figurines*.

19 Laura Battini, 'L'épiphanie divine en Mésopotamie a travers les terres cuites,' *Akkadica* 138 (2017): 69-106.

and means also that the context of discovery has but little significance: it gives us only the final destinations of terracottas but can not explain everything about them. They were moveable and their function is therefore more in relation to their mobility than to the context of discovery. So, aside from style considerations, there is considerable work to be be done on terracottas.

IV. Final remarks

Kassite clay figurines depend on Old Babylonian ones for style and subjects. Of course, the two traditions are not exactly the same: apparent shifts in meaning over time are especially important because they may show an alteration in needs and beliefs, but they are perhaps in part simply due to the chance of excavation. For example, the absence of divinities is not anodyne and it fits well with the Kassite preference for representing divinities by symbols. Other changes in subjects in the Kassite output still require study to be understood, as for example, the introduction of felines, or the female hiding her eyes with her hands. These difficult subjects are a great challenge for understanding the spiritual world and the human needs of ancient people. And indeed, if terracottas have continued to be produced since the 3rd millennium B.C. up into the 1st millennium B.C., it is because they responded very well to the needs of the people of Mesopotamia in all their richness. And this is why archaeologists need to be interested in them.

Table I : Number of Terracottas for each period

1st half of the II mill. BC	**2nd half of the II mill.** BC
South Iraq: Tello= 468 Uruk= 461 Ur= 312 Nippur= 273 (OIP)+ (Legrain) Kish= 225 Asmar= 187 Babylon= 153+ perhaps 5 Isin=149 Ishchali= 115 Khafadjah= 66 Tilool Khettab= 59 Larsa=54 (49 (Louvre)+ 5 (Huot)) Tell ed-Der=32 Sippar : 18 (Scheil)+ (Iraq)+ (Belges) Fara= 11 ND= 87 (Louvre) ND= other museums (reserves, exhibition)	South Iraq : Dur Kurigalzu= 'HUNDREDS' (?) (only 43 published) Nippur= 11 (WC Area), half human, half animals Babylon=3
TOTAL: AT LEAST 2513	**TOTAL:** 58 published+ 'hundreds' from Aqar Quf, never published

Table II : Context of discovery

Context of discovery	1st half of the II mill BC	2nd half of the II mill BC
Houses	Ur, Uruk, Isin, Tell ed-Der, Ishchali, Khafadjé,	Nippur
Temples	Ishchali, Khafadjé	
Tombs	Ur	
Palaces		Aqar Quf
Workshop	Probably Ur,	Aqar Quf (?)
Surface/ antiquary market	More than the half of the total TC found	

Table III: Material, Technique, Typology

Material/Technic	1st half of the II mill BC	2nd half of the II mill BC
Material	Ton	Clay
Technique	Mould ; Hand Made	Hand Made
Plaques	1/3	None
Figurines	2/3	**ALL**
Moulds	Some	None

Table IV: Subjects

Subject	1st half of the II mill BC	2nd half of the II mill BC
Naked frontal female	Yes	Yes
Female with child	Yes	
Female dressed	Yes	
Couple of females	Yes	
Divine pair	Yes	
Divinity sitting	Yes	
Divinity standing	Yes	
Temples' facades	Yes	
Man	Yes	Yes
King	Yes	
Grotesque	Yes	
Humbaba	Yes	
Erotic plaques	Yes	
Chariots	Yes	Yes
Models	Yes	
Animals	Yes	Yes

Through a Glass, Darkly?[1] Sea-Water, Bequests, and the Textualism-Tradition Tension in Islamic Law

Suheil Laher

French mathematician Laplace proffered that the probability of truth of historical information diminishes as time elapses, similar to the decrease in visibility that results from successive panes of glass intervening between an object and its viewer.[2] This article studies the difference that time made to the perception of two issues in Islamic law, as attitudes to ways of knowing transitioned towards a greater privileging of textual transmissions over inherited tradition.

Prominent in the arena of 2nd/8th and 3rd/9th century discussions in Islamic law was a dispute over the relative weight of ḥadīths (reports of things advised or performed by Prophet Muhammad) vis-à-vis living tradition of the community. Muslim traditionists (ḥadīth-folk) stressed an epistemology centered on authentication of the transmission-route (isnād) of ḥadīths. If the report was obtained from credible and morally upright narrators narrating one from the other across generations, going back to the Prophet Muhammad, then this was declared sufficient to authenticate the report, thereby making it admissible for legal and theological evidence.[3] This sometimes caused conflict with followers of earlier legal schools who tended to consider ubiquitous, inherited religious tradition (sunnah) as normative (for matters of sacred law) even in the absence of a ḥadīth text. Al-Shāfiʿī (d. 204/820), eponym of one of the Sunnī legal schools, is widely acknowledged as having been prominent in highlighting and grappling with these tensions, and in championing textualism over (text-less) inherited tradition. Notwithstanding his innovative stance:

1. Shāfiʿī himself cautioned against relying solely on ḥadīth-texts that were not corroborated by practical precedent. Shāfiʿi advised, 'You must follow those ḥadīths that the general public (*ʿāmmah*) recognize, and beware of anomalous ones.'[4] A half-century later, this guardedness has still not disappeared from traditionist circles. Tirmidhī (d. 279/892) states that all the ḥadīths he has included in his *Jāmiʿ* (with the exception of two) have been acted on by at least some of the scholars.[5]

1 Disclaimer: despite what the title might suggest, my talk is not about Paul of Tarsus. The only tenuous connection that could be claimed to Paul is that his phrase may have provided the inspiration for the title of Alice through the Looking Glass, which I will briefly mention at the end of the article.

2 Pierre Simon Marquis of Laplace, *A Philosophical Essay on Probabilities*, translated into English by Frederick Wilson Truscott, (New York: Wiley, 1902), 13

3 In practice, the content of the ḥadīth was also taken into consideration, even though the public stance was often to deny this aspect of verification due to polemic considerations. See: Jonathan A.C. Brown, 'How We Know Early Hadīth Critics Did Matn Criticism and Why It's So Hard to Find,' *Islamic Law and Society* 15.2 (2008): 143-184.

4 Shāfiʿī, *al-Umm* (Beirut: Dār al-Maʿrifah, 1990/1410), 7/358.

5 Almost a millenium later, Sunnī-tending Zaydī Ṣāliḥ al-Maqbali (d. 1106/1695) would claim, in retrospect, that prior to the Golden 3rd/10th century of ḥadīth scholarship, the meaning of *ṣaḥīḥ* (soundly-transmitted) among ḥadīth-folk was 'that which is acted upon.' Contemporary Indian ḥadīth scholar Ḥamzah Malībārī asserts that Islam is a live tradition of recited signs (*āyāt matluwwah*), and that every non-abrogated ḥadīth narrated has been acted upon in the community, and action (*ʿamal*) can promote the epistemological status of a ḥadīth from speculative (*ẓannī*) to certain (*qaṭʿī*). Consecutively-transmitted practice of the religion, he explains, implies that in every generation someone has known and acted on the ḥadīth. But we see that this also raises questions of how the widespread transmission of such practices might or might not be maintained in the future with disintegration of community and secularization of laws.

2. Shāfi'ī also acknowledged the weight of community traditions not backed by a reliably-transmitted text, provided that the practice in question be so ubiquitous in the Muslim community as to constitute a consensus, or be supported by reasoning based on other sacred texts. Despite Shāfi'ī's promotion of soundly-transmitted ḥadīths as a sacred textual source (for theology and law) second only to the Qur'ān, this attitude of accepting inherited practice in matters for which only a weak ḥadīth has been transmitted once again makes its way through his sieve and persists for some decades at least. We can illustratively return once again to Tirmidhī, who cited 'amal (practice within the early religious community) as a justification for pronouncing the call to prayer (adhān) in the ear of the newborn, even though the ḥadīth on this is weak. [6]

The first of these points is not particularly problematic to the methodology of the traditionists, for it does not dispense with isnad-analysis, but rather involves subsequently applying an additional criterion. Not so the second point, for the notion that knowledge with a scripturally-revealed basis could be obtained without recourse to a transmitted text, was nebulous to the ḥadīth-folk, and posed a threat to the entire edifice of their enterprise of textual authentication. Indeed, one of the arguments Shāfi'ī had raised against the 'old' legal schools was that they sometimes accept a practice as religiously normative on the basis of inherited communal practice, without realizing that the practice in question may have been introduced decades after the Prophet Muḥammad's death by a 'secular' government official.[7] Ibn Ḥazm (d. 456/1065), a later scholar who was nevertheless staunchly committed to the methods of the ḥadīth-folk, insisted that the ḥadīth that is the basis of transmitted practice must itself be transmitted, for it is not possible that an action remain in existence while the Prophetic ḥadīth, that enjoined it in the first place, disappear.[8]

The question here is: How (if at all it is possible) can one gauge or pin down that a particular religious practice (or belief) is normative, in the absence of a reliable transmitted text about it? With this in mind, I now turn to two specific issues in Islamic law:

(1) whether a Muslim is allowed to bequeath something to a relative who is already one of the designated heirs under the Qur'ānic laws of inheritance, and
(2) whether it is permissible and valid to use sea-water for ritual ablutions.

For each of these two issues, Shāfi'ī cites a ḥadīth whose isnād does not match up to the traditionalists' criteria for soundness.

(i) Shāfi'i states that the legal scholars and chroniclers before him,[9] 'do not differ that the Prophet said in the Year of the Triumph, 'There is no bequest for an heir,' and they transmit it from those of the people of knowledge from whom they have recorded, so that it is the transmission of public multitude (*'āmmah*) from public multitude, and is

Malībārī, *'Ulūm al-Ḥadīth fī Ḍaw' Taṭbīqāt al-Muḥaddithīn al-Nuqqād* (Beirut : Dār Ibn Ḥazm, 2003), 91-97.

6 Similarly, Imām Aḥmad ibn Hanbal is reported to have approved the practice of *talqin* (reminder to the deceased of the core beliefs by his graveside, to help him answer the angels correctly), despite the ḥadīth about it being weak. Ibn Qayyim al-Jawziyyah, *Kitāb al-Rūḥ* (Beirut: Dar al-Kutub al-'Ilmiyyah, n.d.), 13.

7 See, Abū Yūsuf's criticisms of al-Awzā'ī's appeals to *'amal* in: Abū Yūsuf, *Radd 'alā Siyar al-Awzā'ī* (Hyderabad, Lajnat al-Ma'ārif al-Nu'māniyyah, n.d.).

8 Ibn Ḥazm, *Al-Iḥkām fī Uṣūl al-Aḥkām* (Beirut : Dār al-Afāq al-Jadīdah, 1403/1983), 1/111, 1/214-36.

9 Literally, 'the people of fatwā and those from whom we have recorded of the people of *maghāzī*.

stronger in some cases than the transmission of one [person] from one. Thus have we found the people of knowledge in consensus.'[10]

(ii) The ḥadīth stating that, the sea's water is pure, and its dead [creatures] lawful [for food],'.[11] Shāfi'i quotes the ḥadīth, then comments that its isnād contains an unaccredited narrator (i.e. that it is weak), but that its meaning corresponds to the apparent implication (*ẓāhir*) of the Qur'ān.

For Shāfi'ī, both these ḥadīths lack an isnād that is acceptable according to the traditionists' authentication criteria, but he effectively accepts both ḥadīths, albeit for different reasons. He rules in accordance with the contents of the bequest ḥadīth, because it is clearly knowledge that he considers ubiquitous.[12] So, for Shāfi'ī, ubiquitous knowledge of a practice is the guarantor of its normativity. For the sea-water ḥadīth, in contrast, Shāfi'i makes no such claim of ubiquity; rather he seems to be citing the weak ḥadīth only incidentally, his primary evidentiary reliance being on deduction from a Qur'ānic verse. Written sources confirm that sea-water was not deemed purifying by some prominent Sunnī authorities who significantly pre-date Shāfi'ī.[13]

So, what happened after Shāfi'ī? Firstly, we may note that a number of later traditionists judged the sea-water ḥadīth to be of sound transmission, i.e. they disagreed with Shāfi'ī's dismissal of it as weak.[14] These later authentications can be largely attributed to the uncovering/discovery of additional isnads and further information subsequent to Shāfi'ī. However, these later authentications do not have any bearing on my general thesis here, for Shāfi'ī was working under the assumption of the ḥadīth having a weak isnād. Also, as we will see, at least one prominent 5th/11th century ḥadīth expert agreed with Shāfi'ī's evaluation of this isnād being weak.

The second development that I highlight is that two centuries later, we find a renowned ḥadīth-master Ibn 'Abd al-Barr (d. 463/1071) writing that even though the isnād of the sea-water ḥadīth is not acceptable by ḥadīth-folk's standards,[15] nevertheless it is *ṣaḥīḥ* according to him, because it has been 'received with acceptance' by the scholars. He also states that the scholarly consensus (*ijmā'*) on the meaning of this ḥadīth dispenses with the need for isnād![16] It is somewhat surprising to find Ibn 'Abd al-Barr, who is clearly one of the ḥadīth-folk, overriding isnād-analysis to judge a non-ubiquitous, weak ḥadīth as *ṣaḥīḥ*. The use of the term '*ṣaḥīḥ*' by a ḥadīth-master is typically an indication of confidence that the Prophet Muḥammad uttered words to this effect, a claim that would be difficult to justify here.[17] While Ibn 'Abd al-Barr

10 Shāfi'ī, *Risālah* (Cairo: Ḥalabī, 1940/1358) , 139-140, See also p. 422, where he refers to the Prophet's letter to 'Amr ibn Ḥazm, saying that the scholars accepted it as authentic. The isnāds for the ḥadīth are controversial, even though some have judged some of them to be sound. See the editor, Aḥmad Shākir's comments.

11 Narrated by Mālik in his *Muwaṭṭa'*.

12 among scholars of fiqh and sīrah (but not among scholars of ḥadīth).

13 The sources that transmit this dissent are slightly later than Shāfi'ī: 'Abdul-Razzāq, *Muṣannaf* (Johannesburg: Majlis 'Ilmī, 1983/1403), 1/93; Ibn Abī Shaybah, *Muṣannaf* (Riyad: Maktabat al-Rushd, 1989/1409), 1/122; Tirmidhī, *Jāmi'* (Cairo: Ḥalabī, 1975/1395), 1/100.

14 Bukhārī, when asked about the ḥadīth by Tirmidhī, said it is *ṣaḥīḥ*. Tirmidhī, *Kitab al-Ilal al-Kabir* (Beirut: 'Ālam al-Kutub, 1989/1409), 41. Ibn 'Abd al-Barr strongly disagrees, saying, 'I do not know what this means from Bukhārī. If it had been *ṣaḥīḥ* for him he would have included it in his book.' Others who authenticated the ḥadīth include: Ibn Khuzayma (d. 311), Ibn al-Sakan (d. 353), Ibn Hibban (d. 354), Ibn Mandah (d. 395).

15 Ibn 'Abd al-Barr apparently does not agree with the numerous post-Shāfi'ī figures who did judge the ḥadīth to be *ṣaḥīḥ*.

16 Namarī, *Tamhid* (al-Muḥammadīyah: al-Maṭbaʿah al-Faḍlīyah,1988-), 16/218-219 and *Istidhkār* (Beirut : Dār al-Kutub al-ʿIlmīyah, 2000), 2/95-108. He also observes, while discussing a detail of zakat on cattle, that the ruling is backed by consensus, and established by transmission of *kāffah* from *kāffah*, but individuals (*āḥād*) also transmitted it. *Tamhīd*, 20/139.

17 Unlike for the bequest ḥadīth, see Shāfi'i's comments above.

himself may have been using the term *ṣaḥīḥ* to refer to only to validity of meaning, nevertheless, a slightly earlier legal theoretician Abū Naṣr 'Abdu'l-Wahhāb al-Mālikī (d. 422/1032) is said to have held that a ḥadīth is definitively considered to be *ṣaḥīḥ* if the scholars' 'received it with acceptance.'[18]

Note that what was a weak or disputed ḥadīth is now considered *ṣaḥīḥ* (soundly transmitted) by Ibn 'Abdil-Barr, and also that the early Sunnī dissenters on the sea-water issue (presumably the dissenting view is extinct by his time) have been overridden by a later consensus. It is worthwhile, at this point, to see where the Shī'ites stand with regard to the claims of consensus on sea-water and bequest. The Imāmī Shī'ah agree on the purificatory nature of sea-water. According to 'Āmilī's and Majlisī's encyclopedic compendia, there are numerous narrations to this effect attributed to the Imāms (primarily Ja'far al-Ṣādiq). A ḥadīth attributed to the Prophet, with the same words as the above-mentioned Sunnī narration is mentioned only by al-Muḥaqqiq al-Ḥilli (d. 676/1277), and without an isnād.[19] They also narrate that the Prophet made a statement about bequests to heirs, but that it was in the Farewell Pilgrimage (rather than the Triumph of Makkah), and that it was followed by a qualification: 'There is no bequest for an heir with more than one-third [of the estate].' The Zaydīs have a similar view on both issues.[20] So, even if we concede that a consensus had emerged in the Sunnī world by Shāfi'ī's time, we should note that the dissent remained alive as the dominant view in the Shī'ite world. In other words, even today there is still not true consensus on the matter. Shāfi'ī's claim of consensus must be taken either as referring to a consensus among Sunnīs, or as a near-consensus in which minority dissenters are regarded as inconsequential.[21] Both possibilities are illustrative of how boundaries of orthodoxy contract over time, which is an oft-encountered phenomenon.[22]

So, to recap, Shāfi'ī had accepted text-less tradition only when it was perceived to be so ubiquitous in the Muslim community as to constitute a consensus, or to be supported by reasoning based on other sacred texts. We have seen how the picture blurred with the passage of time, and the two ḥadīths, which al-Shāfi'ī had regarded as having weak transmission, were (some 250 years later) re-classified by Ibn 'Abd al-Barr as sound transmissions backed by consensus. It appears that from the 5th/11th century onward, numerous legal scholars have argued for the ṣaḥīḥ status of several (ostensibly weak) ḥadīths on the basis that the scholars 'received its meaning with acceptance'.[23] By the 10th/16th century, we find Ṣuyūṭī and Sakhāwī writing that the class of acceptable (*maqbūl*) ḥadīths includes those that have been 'received with acceptance' even if they have no isnād. I proffer that this is illustrative of the increasing elusiveness – as generations passed, some earlier views became extinct, and others took on sectarian overtones[24] – of identifying ubiquity and consensus in the knowledge of the

18 Ibn Ḥajar, *Nukat, 1/373, vide* Mahir Yāsin Fahl al-Hiti, *Athar 'Ilal al-ḥadīth fi Ikhtilaf al-Fuqahā', ('Amman: Dar 'Ammār,* 2000), 38.

19 See: Majlisī, *Biḥār al-Anwār* (Beirut: Dār Iḥyā' al-Turāth al-ʿArabī, 1983), 77/108; 'Āmilī, *Wasā'il* (Qum : Maktabat al-Muḥammadī, 1376-1389 [1956-1969]), 2/136; *Mu'tabar* (Qum: Mu'assasat Sayyid al-Shuhadā', 1364/1985), 1/36.

20 *Wasā'il*, 19/290. Hārūnī, *al-Taḥrīr fil-Kashf ʿan Nuṣūṣ al-A'immah al-Naḥārīr*. (Sanʿā': Maktabat Badr, 2012/1433), 23, 402.

21 In the light of general principles or specific evidences.

22 John B. Henderson, *The Construction of Orthodoxy and Heresy*, (Albany: State University of New York Press, 1998).

23 Ibn al-Humām, *Fatḥ al-Qadīr*, 3/143. Ibn Furak (d. 406/1015) and Abu Ishaq al-Isfarayini (d. 418/1027) are perhaps the first to make such a claim (as opposed to merely saying that the 'reception with acceptance' provides sufficient grounds to issue a legal ruling that is in accordance with the weak ḥadīth. See: Mustafa al-Sulaymani, *Itḥāf al-Nabīl bi-Ajwibat as'ilat ʿUlūm al-Ḥadīth : wa-al-ʿIlal wa-al-Jarḥ wa-al-Ta'dīl* ('Ajmān : Maktabat al-Firqān, 2000), 246-250.

24 Dr. 'Abd al-Ghaniyy Muzhir says, in *Usul al-Tashih wal-Tad'if*, that his study of the usage of the term 'the scholars received it with acceptance,' reveals that it is used loosely, to refer to agreement of the majority of scholars, but not a complete consensus. Internet source: http://www.startimes.com/f.aspx?t=33766445, accessed 1/25/18.

earliest Muslims, and that the sunnah-to-ḥadīth shift spearheaded by Shāfiʿī was therefore a natural response to the decay of knowledge over time.

In the 2016 film Alice Through the Looking Glass, Alice returns from the seas to find her father has died, and her ex-suitor Hamish wants her to sign over her ship to him in exchange for the family home. As you know, she refuses, and heads back to sea instead. Shāfiʿī spearheaded a movement to nail down the elusiveness of non-textual tradition into the confines of text (even if it be a text with weak transmission). But the consensus that Shāfiʿī's later ideologues often appealed to was itself quite elusive. In a sense, Shāfiʿī's maiden voyage returned to its home-port, and in the real-life Islamic law version of Alice Through the Looking Glass, Alice marries her long-time suitor, after eventually realizing how much they have always had in common.

Wrinkles in Time

Converging Lines of Evidence: Archaeological and Chronological Data Supporting a 792 BC Date for the Iron IIA-IIB Transition in the Southern Levant

Jeffrey P. Hudon

Ode to a Chronological Chart

Oh we salute thee mighty Chart
Conceived of genius, born of art
Thou subtle scheme of dates veracious
Full of charm and so tenacious
Which with aesthetic charm appealing
Oft evokes such tender feeling
Oh Gem of Truth we now embrace thee
(Knowing none will e'er erase thee)
For knowledge which thou givest freely
Through the courtesy of Edwin Thiele.[1]

Introduction

The city of Lachish has long been recognized as the premier 'type site'[2] for the history and material culture of Judah from the late tenth to the early sixth centuries, BC for primarily two reasons: Lachish served as an important administrative city for Judah, surpassed in significance and size only by Jerusalem during the last two centuries of the monarchy. Lachish perhaps even functioned as Judah's second capital, due to its large 30-acre size, strategic location near Judah's southwestern frontier, sophisticated fortifications and imposing Palace-Fort complex as well as serving as Amaziah's final redoubt following a Jerusalem coup.[3] While Jerusalem remained continually inhabited during the ensuing centuries, which resulted in poorly preserved remains from its earlier periods, the Iron Age II occupational levels at Lachish are reasonably well preserved and provide a fairly complete progression of typical Judahite pottery

1 Mr. Albert D. Smouse, while studying at Emmanuel Missionary College (later Andrews University), composed this poem and dedicated it to his religion professor, Edwin R. Thiele in 1954-55. I wish to thank Mr. Smouse for sharing his poems and allowing me to publish one of them here. All dates follow the biblical chronology developed by Thiele (discussed below). I also wish to thank my former teacher and now colleague Paul J. Ray for offering valuable input and feedback. I remain entirely responsible for the content in this study.

2 By the time Orna Zimhoni's comparative study of the Iron Age II pottery from several sites in Judah, 'The Iron Age Pottery of Tel 'Eton and its Relation to the Lachish, Tell Beit Mirsim and Arad Assemblages' *Tel Aviv* 12 (1985): 63-90; and eadem, *Studies in the Iron Age Pottery of Israel, Typological, Archaeological and Chronological Aspects* (Journal of the Institute of Archaeology of Tel Aviv University Occasional Publications 2; Tel Aviv: Tel Aviv University 1997): 179-210, appeared, Lachish had replaced Tell Beit Mirsim for having the distinction as the ceramic type site of Judah. More recently, the rich destruction layer currently under excavation and labeled as stratum A3 at Tell es-Safi (Philistine Gath) has catapulted that site to the forefront of ninth century B.C. ceramic typological studies for the region. For example, see the recent discussion by David Ussishkin, 'Gath, Lachish and Jerusalem in the 9th Cent. B.C.E.-an Archaeological Reassessment,' *Zeitschrift des Deutschen Palästina-Vereins* 131 (2015): 129-49 and references there. Nevertheless, Beth Shemesh Level 3 provides a more distinctively Judahite material cultural corpus of finds relating to this century.

3 e.g., Avraham Faust, *The Archaeology of Israelite Society in Iron Age II* (Translated by R. Ludlum from Hebrew; Winona Lake: Eisenbrauns, 2012), 78-82; William G. Dever, *Beyond the Texts: An Archaeological Portrait of Ancient Israel and Judah* (Atlanta: SBL, 2017), 422-24; 2 Kgs 14:19; 2 Chr 25:27.

typologies, material culture and fortifications utilized throughout the period of the divided monarchy (930-586 BC).[4] Furthermore, Lachish has been extensively excavated, and its data well recorded.[5] Consequently, the Lachish ceramic and material cultural assemblages provide a suitable benchmark for determining whether Judahite occupation and/or hegemony existed over a given site and the surrounding hinterland when based upon the material remains.

The Iron Age II stratigraphy of Lachish also achieved great importance amongst historians and biblical scholars when Israeli scholars, following earlier suggestions made by Olga Tufnell and Richard D. Barnett, decisively correlated two archaeological destruction layers and their associated remains with events recorded in biblical and other historical sources. Specifically, data from Lachish Levels III and II enabled correlations with two recorded attacks against Judah; the Assyrian campaign in 701 BC directed by Sennacherib and the Babylonian onslaught of 587-6 BC led by Nebuchadnezzar.[6] Yet, the dates and agents responsible for the collapse of two earlier Iron Age II occupational strata at Lachish (Levels V and IV) remain obscure and over the past forty years, numerous scholars proposed several historical circumstances, summarized below.[7] Recently, David Ussishkin, who directed excavation and restoration work at Lachish

4 The dates follow the chronology developed and published by Edwin R. Thiele, 'The Chronology of the Kings of Judah and Israel,' *Journal of Near Eastern Studies* 3 (1944): 152-54; 'Pekah to Hezekiah and the Azariah and Hezekiah Synchronisms,' *Vetus Testamentum* 16 (1966): 83-107; *A Chronology of the Hebrew Kings* (Grand Rapids: Zondervan, 1977); 'Two New Assyrian Synchronisms with Jehoash and Menahem,' in *The Archaeology of Jordan and Other Studies Presented to Siegfried H. Horn* (eds. Lawrence T. Geraty and Larry G. Herr; Berrien Springs: Andrews University, 1986), 601-619; and most notably his thrice published *magnum opus, The Mysterious Numbers of the Hebrew Kings: A Reconstruction of the Chronology of the Kingdoms of Israel and Judah* (first edition, Chicago: University of Chicago, 1951; second edition, Grand Rapids: Eerdmans, 1965; third edition, Grand Rapids: Zondervan, 1983). As the ensuing discussion explains, the somewhat limited exposure and subsequently low yields of pottery previously collected from Lachish Levels V-IV are now greatly augmented by the ceramic corpus unearthed by the recent Hebrew University and Southern Adventist University sponsored excavations at the site.

5 For the excavations, see the following final reports: Olga Tufnell, *Lachish III: The Iron Age.* Text and Plates volumes (London: Oxford University, 1953); Yohanan Aharoni, *Investigations at Lachish: The Sanctuary and the Residency (Lachish V)* (Tel Aviv: Gateway, 1975); David Ussishkin, *The Renewed Archaeological Excavations at Lachish (1973-1994)* 5 volumes (Tel Aviv: Emery and Claire Yass Publications in Archaeology, 2004). The main results have recently been summarized and reassessed by Ussishkin, *Biblical Lachish: A Tale of Construction, Destruction, Excavation and Restoration* (trans. M. Feinberg Vamosh from Hebrew; Jerusalem and Washington DC: Israel Exploration Society and Biblical Archaeology Society, 2014). Two teams of scholars recently completed additional excavations the site from 2014-2017; one group focusing on the northeast part of the tel and the other working in the six chamber Iron Age II gateway. On the former, see further below, note 7.

6 In her publications on Lachish, Olga Tufnell was the first scholar to correctly recognize and attribute the dates for two destruction layers to the invasions of Sennacherib in 701 BC and Nebuchadnezzar during 587-86 BC, 'Excavations at Tell Ed-Duweir, Palestine, Directed by the Late J. L. Starkey, 1932-1938: Some Results and Reflections,' *Palestine Exploration Quarterly* 82 (1950): 74-79; *Lachish III*, 47-50, 53-56, 72-74; 'Hazor, Samaria and Lachish: A Synthesis,' *Palestine Exploration Quarterly* 91 (1959): 93-100. Aside from Richard D. Barnett, The Siege of Lachish, Israel, *Israel Exploration Journal* 8 (1958): 161-62; Tufnell's historical conclusions were disputed by most of her colleagues until the renewed excavations by Ussishkin fully verified and exonerated her views, 'Royal Judean Storage Jars and Private Seal Impressions,' *Bulletin of the American Schools of Oriental Research* 223 (1976): 1-3, 11-12; 'The Destruction of Lachish by Sennacherib and the Dating of the Royal Judean Storage Jars,' *Tel Aviv* 4 (1977): 28-54; Reassessment of the Stratigraphy and Chronology of Archaeological Sites in Judah in the Light of Lachish III, in *Biblical Archaeology Today: Proceedings of the International Congress on Biblical Archaeology. Jerusalem, April 1984*, (ed. J. Amitai; Jerusalem: Israel Exploration Society, 1985), 142-144. Two additional landmark studies from this period, one by Anson F. Rainey on historical-geographical issues ('The Fate of Lachish during the Campaigns of Sennacherib and Nebuchadrezzar,' in *Investigations at Lachish: The Sanctuary and the Residency (Lachish V)* [ed. Y. Aharoni; Tel Aviv: Gateway, 1975], 47-60); and that of Yohanan and Miriam Aharoni regarding ceramic evidence (The Stratification of Judahite Sites in the 8th and 7th Centuries BCE, *Bulletin of the American Schools of Oriental Research* 224 [1976]: 73-90) provided crucial additional evidence supporting Tufnell's conclusions.

7 As reported in academic papers delivered at professional meetings and in a public lecture by Michael Hasel ('David and Solomon's Legacy: The Results of the Fourth Expedition to Lachish.' Public lecture delivered at Andrews University, October 16, 2017), the recently concluded excavations at Lachish, conducted by Hebrew University and Southern Adventist University and led by Yosef Garfinkel, Michael G. Hasel, Martin G. Klingbeil and others, unearthed spectacular evidence relating to Levels IV and V along the shoulder and slope of the northeastern sector of the mound, including strong fortifications with attached dwellings and other structures. These new discoveries all but confirm

for over twenty years, made a significant chronological adjustment to his reckoning of the transition date between Lachish Levels IV-III. He now suggests a date at or near 800 BC.[8] This new position regarding dating the end of Lachish Level IV follows similar proposals voiced by several other scholars for the same relative dating and warrants further discussion. Ussishkin's chronological shift, based on ceramic evidence, offers a remarkable archaeological convergence with a progression of events overtly recorded or implied in two biblical texts and also presents an important correlation with the absolute chronology established years ago by Edwin R. Thiele.

This study submits that Lachish Level IV and Beth Shemesh Level 3 both suffered destructive acts in 792 BC directly resulting from Judah's military debacle at Beth Shemesh and Amaziah's capture. Thus, the destructive marks of sacking and pillaging that these two cities display, reflect operations ordered by Jehoash and carried out by his victorious Israelite soldiers. Such activity appears to complement and receive support from the biblical accounts of Jehoash's sacking and plundering of Jerusalem. This proposal, coupled with the extensive archaeological data provided by Tell es-Safi Stratum A3 destruction currently attributed to Hazael, offers historical evidence towards establishing 792 BC as a suitable chronological dividing point between the Iron Age IIA-B and thus providing a more precise historical and archaeological benchmark for the history of Judah and the southern Levant.

The Debate over the Transition Date from Lachish Level IV to Level III

While Sennacherib's violent destruction of Lachish Level III is now confidently dated to 701 BC,[9] the date of establishment, as well as the full chronological span of this stratum remains uncertain, as are the founding and ending dates of the previous two strata, Levels V and IV. While the current paucity of published data consigns the chronology of Lachish Level V beyond the scope of this paper, the differing views regarding when to date the end of Lachish Level IV and the establishment of Lachish Level III discussed below remain one of the unsettled questions concerning the history of Iron Age II period in the southern Levant. These disputed dates rightly serve as a chronologically ambiguous demarcation line separating the Iron Age IIA from the Iron Age IIB period for selected scholars,[10] demonstrating that a wider implication may exist for this chronological question rather than merely determining the date and circumstances of one or more related events. Moreover, establishing the end of Lachish Level IV and the rise of Lachish Level III would provide a better stratigraphic picture of Judah during

---

that Lachish was a large, fortified city during the tenth century, BC and that this Stratum (Level V) appears to have met a violent end. Until detailed scientific reports appear, see the provisional observations of Hoo-Goo Kang ('Four Notes on Tel Lachish Level V,' in *From Sha'ar Hagolan to Shaarim: Essays in Honor of Prof. Yosef Garfinkel* [eds. Saʿar Ganor; I. Kreimerman; K. Streit; and M. Mumcuoglu; Jerusalem: Israel Exploration Society, 2016], 283-94) anticipated also by Ron E. Tappy ('East of Ashkelon: The Setting and Settling of the Judean Lowlands in the Iron Age IIA Period,' in *Exploring the Longue Durée: Essays in Honor of Lawrence E. Stager* [ed. J. David Schloen; Winona Lake: Eisenbrauns, 2009], 456-57), who argues for two phases to Stratum V.

8 Ussishkin, *Biblical Lachish*. See also the author's review of Ussishkin's book (*Andrews University Seminary Studies* 53 (2015): 409-13), which briefly summarizes the conclusions discussed in detail here. Noting that any attempt to establish the termination date of Lachish Level IV is difficult, Ussishkin offers evidence that seems to point towards ca. 800 BC, but stops short of affirming it in a newly published paper, 'Gath, Lachish and Jerusalem in the 9th Cent. BCE.,' (2015) An Archaeological Reassessment,' *Zeitschrift des deutschen Palästina-Vereins* 131 (2015): 137-38, 145.

9 Edward Lipiński, *On the Skirts of Canaan in the Iron Age: Historical and Topographical Researches* (Orientalia Lovaniensia Analecta 153; Leuven: Peeters, 2006), 415-416, represents perhaps the lone exception among current scholars. He assigns Lachish Level III to the seventh century BC.

10 For the resulting historical ambiguity surrounding the transition from Iron Age IIA to Iron Age IIB see the discussion by Ze'ev Herzog and Lilly Singer-Avitz, 'Redefining the Centre: The Emergence of State in Judah' *Tel Aviv* 31 (2004): 228-31 and Table 1.

the tenth and ninth century BC and perhaps shed additional light on the Iron Age IIA high versus low chronological debate.

Dating the Demise of Lachish Level IV

Until recently, a large part of the published archaeological data for Lachish Level IV derived from Area S, the trench excavated by Ussishkin and Barkay directly north of the six-chamber gate complex. According to Ussishkin[11] most of the structures of Level IV, aside perhaps from the city walls, were partially demolished or destroyed suddenly, but then either rebuilt or entirely replaced. Adjacent to Area S, Palace-Fort B was also destroyed, but its foundations were reused and enlarged when Palace-Fort C was erected during the building of Level III. The southern annex, a building complex related to the Palace-Fort B, was also replaced with a similar, but twice as large building. The enclosure wall was rebuilt along the same course and the city walls continued to be used largely without change, but the gate complex received significant modifications. The demolished Level IV domestic structures were rebuilt with a similar plan. Significantly, the excavators recognized that material-cultural continuity between Level IV to Level III was clearly evident, due also in part to the prompt rebuilding of the city. No signs of fire or intentional destruction were found in Area S or elsewhere at the site.[12] Still, Ussishkin's limited exposure of Level IV to date renders this supposition tentative.

Opinions differ regarding the circumstances behind the end of Lachish Level IV, as reflected in the relevant literature over the last 65 years. Prior to Ussishkin's preliminary reports, most scholars, notably Tufnell,[13] Wright[14] and Aharoni and Amiran[15] chose to leave the question open due to the relative paucity of data available from Lachish at that time. Yet, following their view that a comprehensive Babylonian campaign against Judah took place in 597 BC, Albright, Wright and other similar minded scholars should have ascribed the termination of Level IV, at least provisionally, to Sennacherib, but a review of the relevant literature failed to find this position expressed in their publications.

Historical Candidates for the Lachish Level IV Destruction

The Campaign of Hazael (Late Ninth Century BC)

Based solely on the ceramic evidence, Zimhoni[16] argued that a transitional assemblage that contains some earlier (Level IV) and some later (Level III) forms were found at Tel 'Eton and only provided a very broad provisional date of 850-750 BC for Level IV. While she noted that the date and circumstances of Lachish Level IV's demise was not completely certain, Zimhoni[17]

11 *Renewed Archaeological Excavations*, 82-83; Gabriel Barkay and David Ussishkin, 'Area S: The Iron Age Strata,' in vol. 2 of *The Renewed Archaeological Excavations at Lachish (1973-1994)* vol. 2 (ed. David Ussishkin; Tel Aviv: Emery and Claire Yass Publications in Archaeology, 2004): 447; Ussishkin, *Biblical Lachish*, 204-15.

12 Zimhoni, *Iron Age Pottery*, 200; Ussishkin, *Renewed Archaeological Excavations*, 83; *Biblical Lachish*, 214-15)

13 *Lachish III*, 53-55.

14 G. Ernest Wright, 'Judean Lachish,' *Biblical Archaeologist* 18 (1955): 12; eidem, Review of *Lachish III: The Iron Age*, by O. Tufnell, *Journal of Near Eastern Studies* 14 (1955): 188; idem, Review of *Lachish III: The Iron Age*, by O. Tufnell, *Vetus Testamentum* 5 (1955): 99.

15 Yohanan Aharoni and Ruth Amiran, 'A New Scheme for the Sub-Division of the Iron Age in Palestine,' *Israel Exploration Journal* 8 (1958): 183.

16 *Iron Age Pottery*, 173, 207-8.

17 *Iron Age Pottery*, 171-73, 200; 'The Pottery of Levels V and IV and its Archaeological and Chronological Implications,' in *The Renewed Archaeological Excavations at Lachish (1973-1994)* (ed. David Ussishkin. Tel Aviv: Emery and Claire Yass Publications in Archaeology, 2004), 1643-45, 1706-7.

nevertheless concluded: 'that the city of Level IV flourished and came to an end sometime during the 9th century. In spite of the lack of precise dating of this level, its well-stratified assemblages *which differ substantially from those of Level III* are useful for comparisons with material at Lachish itself (Level V) and other sites' (italics mine). Based upon parallels from Tel Batash, Mazar and Panitz-Cohen[18] view a date during the second half of the ninth century BC for the termination of Lachish Level IV.

The Campaign of Hazael (e.g., 2 Kgs 12:17-18; 13:22) provides the most sensible historical context for this dating. Based upon his acceptance of a late ninth century BC destruction date for Tell es-Safi Stratum A3, and acknowledging that a close resemblance exists between the Safi A3 pottery to the Lachish Level IV assemblage, Ussishkin[19] changed his dating of the Lachish Level IV pottery to ca. 800 BC Ussishkin rejects any argument that the two cities fell simultaneously. Again, based upon ceramic comparisons between the two assemblages and their respective historical contexts, he argues that Lachish IV continued to exist for a short time after Hazael destroyed Tell es-Safi-Gath.[20] One outstanding question concerning the Safi A3 destruction concerns the as yet undetermined date of Hazael's campaign. Scholars offer various suggestions spanning more than a quarter century, which do not appreciably narrow the range of dates to be useful for inclusion here.

A Destruction dating to ca. 800 BC

A date near or at 800 BC was proposed by Aharoni,[21] who suggested that Lachish IV terminated around the end of the ninth or early in the eighth century BC Aharoni also recognized ceramic parallels from Arad Stratum X and Beth-Shemesh Stratum IIb (using the Haverford excavation terminology, which parallels Level 3 of the new excavations). Several other scholars have followed this date, including Y. Shiloh, Herzog and Singer-Avitz, A. Fantalkin and I. Finkelstein, Na'aman,[22] S. Bunimovitz and Z. Lederman (after an assessment of the Beth Shemesh material),[23] A. Maeir[24] (based upon the late Iron Age IIA pottery found in the destruction of Tell es-Safi-Gath Stratum A3), and Ussishkin.[25] However, none of the scholars have, to the best of my knowledge,

18 *Timnah (Tel Batash) II: The Finds from the First Millennium B.C.E. Text.* (Qedem 42; Jerusalem: Institute of Archaeology, Hebrew University), 275.

19 'On the So-called Aramaean 'Siege Trench' in Tell es-Safi, Ancient Gath,' *Israel Exploration Journal* 59 (2009): 140; *Biblical Lachish,* 16, 212; 'Gath, Lachish and Jerusalem,' 138.

20 See the important observations by Itzhaq Shai and Aren M. Maeir ('The Late Iron Age IIA Pottery Assemblage from Stratum A3,' in *Tell es-Safi/Gath I: The 1996-2005 Seasons Volume 1: Text* [ed. Aren M. Maeir, Ägypten und Altes Testament 69; Wiesbaden: Harrassowitz, 2012]: 345) noting the differences between Lachish IV-III pottery and the rich Tell es-Safi (Gath of the Philistines) Stratum A3 assemblage.

21 *Investigations at Lachish,* 14-15, 45.

22 Yigal Shiloh, 'Judah and Jerusalem in the Eighth-Sixth Centuries B.C.E.' in *Recent Excavations in Israel: Studies in Iron Age Archaeology* (eds. Seymour Gitin and William G. Dever; Annual of the American Schools of Oriental Research 49; Winona Lake: Eisenbrauns, 1989) 103, Fig. 5.5; Ze'ev Herzog and Lilly Singer-Avitz, 'Redefining the Centre: The Emergence of State in Judah,' *Tel Aviv* 31 (2004): 229-31, 236; Alexander Fantalkin and Israel Finkelstein, 'The Sheshonq I Campaign and the 8th-Century B.C.E. Earthquake—More on the Archaeology and History of the South in the Iron I-IIA,' *Tel Aviv* 33 (2006): 24); Nadav Na'aman, 'When and How Did Jerusalem Become a Great City? The Rise of Jerusalem as Judah's Premier City in the Eighth-Seventh Centuries B.C.E.,' *Bulletin of the American Schools of Oriental Research* 347 (2007): 25.

23 Shlomo Bunimovitz and Zvi Lederman, 'Close Yet Apart: Diverse Cultural Dynamics at Iron Age Beth-Shemesh and Lachish,' in *The Fire Signals of Lachish: Studies in the Archaeology and History of Israel in the Late Bronze Age, Iron Age, and Persian Period in Honor of David Ussishkin* (eds. Israel Finkelstein and Nadav Na'aman; Winona Lake: Eisenbrauns, 2011), 45.

24 Aren Maeir, 'The Tell es-Safi/Gath Archaeological Project 1996-2010: Introduction, Overview and Synopsis of Results,' in *Tell es-Safi/Gath I: The 1996-2005 Seasons Volume 1: Text* (ed. Aren M. Maeir; Ägypten und Altes Testament 69; Wiesbaden: Harrassowitz, 2012), 38-39.

25 *Biblical Lachish,* 16, 214-15.

attempted to connect the destruction to any historical event or suggested a specific agent responsible for this destruction. Ussishkin[26] in particular considered possible agents, but admits: 'it is difficult to decide among the various possibilities.'

Actions by Jehoash of Israel (792 BC)

Events related to the defeat of Amaziah at Beth Shemesh, specifically an unrecorded Israelite sack of Lachish by Jehoash, provides an attractive historical setting for the end of Level IV, as I already suggested elsewhere.[27] The date, which falls at the beginning of the eighth century BC, correlates with the ceramic evidence and Ussishkin's belief that Lachish IV continued for a short time after Tell es-Safi A3. Moreover, Jehoash broke down about 400 cubits (600 feet) of Jerusalem's wall, yet the biblical account makes no mention of his burning Judah's capital (2 Kgs 14:14-15; 2 Chr 25:23-24). Considering that Lachish functioned as an important administrative city for Judah at that time, it is difficult to imagine that Jehoash entirely ignored Lachish, failing to carry out arbitrary destructive acts there as well in order to further humiliate and disgrace Amaziah and his population and weaken Judah's military capabilities. Naturally, the key evidence rests upon an extensive comparison between the pottery from Lachish IV and the ceramic assemblage of Beth Shemesh Level 3.[28] Accordingly, based primarily on ceramic evidence, the excavators of Beth Shemesh repeatedly claim that Level 3 exhibits an early eighth century BC destruction, most likely attributed to Jehoash in ca. 790 BC.[29] Following their conclusions published in an earlier summary report,[30] Finkelstein[31] also supports a late ninth-early eighth century BC date for the destruction of Beth Shemesh.

The Death of Amaziah (767 BC)

Events surrounding the murder of Amaziah (2 Kgs 14:19; 2 Chr 25:27), proposed by Aharoni[32] and also advocated earlier by this writer[33] and Na'aman,[34] provides yet another historical background for the fall of Lachish IV. Whereas Na'aman dates this event to 788 BC, the year of Amaziah's death was 767 BC, according to Thiele's (1965: 77-79, 205; 1977: 40-45, 75; 1983: 113-23, 217) chronology, and marked the end of a twenty-four-year co-regency with Uzziah. While Ussishkin (1993: 907; 1997: 320; 2004: 83) demonstrates uncertainty as to whether Amaziah's

---

26 Ibid., 215.

27 Review of *Biblical Lachish: A Tale of Construction, Destruction, Excavation and Restoration,* by D. Ussishkin, *Andrews University Seminary Studies* 53 (2015): 412.

28 As noted earlier, Aharoni (*Investigations at Lachish,* 14-15) recognized similarities in the ceramics from Lachish IV and Beth Shemesh (Stratum 3) forty years ago. The final report for the Beth Shemesh Iron Age remains appeared only after this paper was written, but a brief examination of these volumes failed to reveal any historical correlations between the destructions of Beth Shemesh Stratum 3 and Lachish Level IV pointed out by the authors, see Shlomo Bunimovitz and Zvi Lederman, *Tel Beth-Shemesh A Border Community in Judah. Renewed Excavations 1990-2000: The Iron Age.* 2 Volumes (Tel Aviv and Winona Lake: Emery and Claire Yass Publications in Archaeology and Eisenbrauns, 2016), 357-70, 379-82.

29 Shlomo Bunimovitz and Zvi Lederman, 'The Archaeology of Border Communities. Renewed Excavations at Tel Beth-Shemesh, Part 1: The Iron Age,' *Near Eastern Archaeology* 72 (2009): 136; eidem, 'Close Yet Apart,' 44-45; eidem, *Tel Beth-Shemesh*, 363-65, 382; eidem, 'Swinging on the 'Sorek Seesaw': Tel Beth-Shemesh and the Sorek Valley in the Iron Age,' in *The Shephelah during the Iron Age: Recent Archaeological Studies* (eds. Oded Lipschits and Aren M. Maeir; Winona Lake: Eisenbrauns, 2017), 34.

30 Shlomo Bunimovitz and Zvi Lederman, 'The Iron Age Fortifications of Tel Beth Shemesh: A 1990-2000 Perspective,' *Israel Exploration Journal* 51 (2001): 144.

31 Israel Finkelstein, 'Chronology Rejoinders,' *Palestine Exploration Quarterly* 134 (2002): 121-24.

32 *Investigations at Lachish*, 41.

33 J.P. Hudon, 'The LMLK Storage Jars and the Reign of Uzziah: Towards a mid-eighth century B.C. *terminus a quo* for the Royal Jars of the Kingdom of Judah,' *Near East Archaeological Society Bulletin* 55 (2010): 33-34.

34 Nadav Na'aman, 'The Kingdom of Judah in the 9th Century BCE: Text Analysis versus Archaeological Research,' *Tel Aviv* 40 (2013): 253.

death occurred during Level IV or Level III, he does postulate that Amaziah took refuge inside the Palace-Fort and was probably murdered there (Ussishkin 1978b: 22-23: 2004: 83). Nevertheless, this late date is difficult to maintain in light of the more recent ceramic evidence cited above. Furthermore, while Amaziah's rebellion and death may have caused damage to the city, it seems highly improbable that Judahite soldiers, aside from severely punishing rebellious elements in the city loyal to Amaziah, would destroy the Palace-Fort and defensive walls of Judah's second most important city, even if a substantial contingent of soldiers loyal to the king temporarily controlled it.

The Earthquake during the days of Uzziah (ca. 760 BC)

The earthquake during Uzziah's reign (Amos 1:1; Zech. 14:5) is difficult to precisely date and theoretically could fall anywhere between 792-740 BC However, many scholars posit a date of ca. 760 BC. First suggested orally by M. Kochavi during a visit to Lachish in 1976, the theory cautiously accepted by some and nearly uncritically adopted by others.[35] Zimhoni[36] remarked that: 'Acceptance or rejection of this date is most significant in determining the beginning of the development of the vessel assemblage, which was found, broken and burnt, upon the floors of destroyed Level III structures. In this developed and unified assemblage, there are no remains of any vessels which were in use during Level IV. Therefore, the logical period of its development would be during the second half of the 8th century.' Zimhoni's words stands in marked contrast to her observation that Lachish IV ended during the ninth century BC. Kochavi's earthquake hypothesis, once an attractive proposal, has long been disputed by

35 e.g., Ussishkin, 'Destruction of Lachish,' 52; idem, 'Lachish,' in *The New Encyclopedia of Archaeological Excavations in the Holy Land* 3 (ed. E. Stern; New York: Simon and Schuster, 1993), 907; idem, 'Lachish,' in *The Oxford Encyclopedia of Archaeology in the Near East* 3 (ed. E. M. Meyers; New York: Oxford University, 1997), 320; idem, *Renewed*, 83; idem, *Biblical Lachish*, 212-15, 223; Ze'ev Herzog, 'The Fortress Mound at Tel Arad: An Interim Report,' *Tel Aviv* 29 (2002): 96-98; Lilly Singer-Avitz, 'Arad: The Iron Age Pottery Assemblages,' *Tel Aviv* 29 (2002): 162-80; Herzog and Singer-Avitz, 'Redefining the Centre', 229-31; Gabriel Barkay, The Iron Age II-III. in *The Archaeology of Ancient Israel* (ed. Amnon Ben-Tor; trans. R. Greenberg: New Haven: Yale 1992), 345; Barkay and Ussishkin, 'Area S,' 447; Zimhoni, *Iron Age Pottery*, 172-73; eadem, 'Pottery of Levels V and IV,' 1707; Dever, *Beyond the Texts*, 424. The literature devoted to this earthquake and its place in the archaeological and historical record has grown vastly in recent years. See the following representative publications: Amotz Agnon, 'Pre-Instrumental Earthquakes along the Dead Sea Rift.' in *Dead Sea Transform Fault System: Reviews* (eds. Zvi Garfinkel; Zvi Ben-Avraham; and Elisa Kagan. Modern Approaches to Solid Earth Sciences 6. Dordrecht: Springer, 2014): 234-38; Nicholas N. Ambraseys, *Earthquakes in the Mediterranean and Middle East: A Multidisciplinary Study of Seismicity up to 1900* (Cambridge: Cambridge University, 2009): 68-78; S.A. Austin, G.W.Franz and E. G. Frost, 'Amos's Earthquake: An Extraordinary Middle East Seismic Event of 750 B.C.,' *International Geology Review* 12 (2000): 657-71; K. J. Dell, 'Amos and the Earthquake: Judgment as Natural Disaster' in *Aspects of Amos: Exegesis and Interpretation* (eds. Anselm C. Hagedorn and Andrew Mein; Library of Hebrew Bible/Old Testament Studies 536; New York: T & T Clark, 2011), 1-14; William Dever, 'A Case-Study in Biblical Archaeology: The Earthquake of *CA.* 760 B.C.E.,' *Eretz-Israel* 23 (1992): 27*-35*; Diana V. Edelman, 'Earthquakes in the Ancient Southern Levant: A Literary Topos and a Problem Requiring Architectural Solutions,' in *Disaster and Relief Management* (ed. A. Berlejung; Forschungen zum Alten Testament 81; Tübingen: Mohr Siebeck, 2012), 205-38; Fantalkin and Finkelstein, 'Sheshonq,' 18-42; G. Franz, *The Earthquake in the Days of King Uzziah: An Archaeological, Geological and Theological Appraisal.* (Tacoma: Evangelical Theological Society, 1990); David Noah Freedman and A. Welch, 'Amos's Earthquake and Israelite Prophecy, in *Scripture and Other Artifacts: Essays on the Bible and Archaeology in Honor of Philip J. King* (eds. Michael D. Coogan, J. Cheryl Exum, and Lawrence E. Stager; Louisville: Westminster/John Knox, 1994), 188-98; Michael G. Hasel, 'New Early Eighth-Century B.C. Earthquake Evidence at Tel Gezer: Archaeological, Geological, and Literary Indications and Correlations' (Unpublished M.A. Thesis, Andrews University, 1992); Hudon, 'LMLK,' 36; Aren M. Maeir, 'Philistia and the Judean Shephelah after Hazael and the 'Uzziah Earthquake: The Power Play between the Philistines, Judahites and Assyrians in the 8th Century BCE in Light of the Excavations at Tell es-Safi/Gath,' in *Disaster and Relief Management* (ed. Angelika Berlejung; Forschungen zum Alten Testament 81; Tübingen: Mohr Siebeck. 2012), 241-62; Amos Nur and Dawn Burgess, *Apocalypse: Earthquakes, Archaeology, and the Wrath of God* (Princeton: Princeton University, 2008), 192-194; D. K. Ogden, 'The Earthquake Motif in the Book of Amos,' in *Goldene Äpfel in silbernen Schalen: Collected Communications to the XIIIth Congress of the International Organization for the Study of the Old Testament, Leuven 1989* (eds. Klaus-Dietrich Schunck and Matthias Augustin; Beiträge zur Erforschung des Alten Testaments und des Antiken Judentums 20; Frankfurt am Main: Peter Lang, 1992), 69-80.

36 *Iron Age Pottery*, 173.

many scholars,[37] and now rests on unstable foundations.[38] The substantial differences from the Level IV pottery to that of Level III, coupled with the complete absence of any telltale signs of seismic activity at the site,[39] serves to further weaken this view. Indeed, according to Thiele's chronology, a date of 800 BC coincides just before the beginning of Uzziah's co-regency with his father, while a date around 760 BC falls far into the second half of his rule.

The Founding Date of Level III

Over the years, a number of scholars have either argued, or at least recognized that Lachish Level III possibly existed for about a century.[40] This date also seems to fit the current majority view concerning the chronology of Level III. For example, Fantalkin and Finkelstein[41] posit that close similarities between the Iron Age IIB pottery of Arad stratum X, IX and VIII and Lachish Level III demonstrate that the latter must have been founded very early in the eighth century BC. Furthermore, the strength of the new Level III (not Level IV as argued by Ussishkin)[42] defenses and most likely a fortified palace from which to exercise his power were probably among the factors that enticed Amaziah to flee from Jerusalem to Lachish. Bunimovitz and Lederman[43] conclude that Level III at Lachish should represent a long-lasting settlement, which stood for nearly a century, since its typical pottery assemblage seems to appear already at Beth-Shemesh in the early eighth century BC. In his recent popular summary treatment of Lachish, Ussishkin[44] acknowledged that ca. 800 BC is a preferable date for the establishment of Level III. Thus, if the destruction of Lachish Level IV occurred in 792 BC following Thiele's chronology, the construction of Level III must have commenced very early during the co-regency of Amaziah and Uzziah (ca. 791-790 BC). Uzziah certainly initiated this rebuilding project in the wake of his father's defeat and temporary exile to Israel.[45] If the city was violated and extensively damaged after the debacle at Beth Shemesh, Judah's greatly weakened geopolitical status would have demanded the repair and restoration of Lachish, as well as Jerusalem, as quickly as possible. As noted above, evidence clearly indicates that Lachish was resettled by Judahites, based upon close similarities in pottery, domestic architecture and other aspects of their material culture. For example, Zimhoni[46] states that '(t)he basic plan of the Level III houses closely resembles that of Level IV. Some of the earlier walls remained unchanged, while strengthening walls were added to others. In some cases, internal spaces were divided differently. Level III floors were extremely close—about 10-20 cm.—to those of Level IV, and in the course of the

37 e.g. Mazar and Panitz-Cohen, *Timnah*, 274-75; Anson Rainey, 'Stones for Bread: Archaeology versus History,' *Near Eastern Archaeology* 64 (2001): 142; Avraham Faust, 'The Settlement of Jerusalem's Western Hill and the City's Status in Iron Age II Revisited,' *Zeitschrift des deutschen Palästina-Vereins* 121 (2005): 106-7 and n. 13; Fantalkin and Finkelstein, 'Sheshonq,' 22-24; Hudon, 'LMLK,' 33; Na'aman, 'Kingdom of Judah,' 252-55; Bunimovitz and Lederman, '*Tel Beth-Shemesh,*' 363-65, 382.

38 Bunimovitz and Lederman ('Close Yet Apart,' 43-45; eidem, *Tel Beth-Shemesh*, 363-65) discuss the ideological reasons for the popularity of this position, notably for 'scholars seeking a historical datum line for the transition from Iron Age IIA to Iron Age IIB in Judah... and seems to be a factoid constructed by current archaeological research rather than past reality.'

39 e.g. Bunimovitz and Lederman, 'Border Communities', 136; Ussishkin, 'Gath, Lachish and Jerusalem,' 138.

40 e.g. Shiloh, 'Judah and Jerusalem,' 103, Fig. 5.5 (dates Level III to ca. 800 B.C.); Mazar and Panitz-Cohen, *Timnah*, 274-75; Philip J. King, 'Why Lachish Matters: A Major Site Gets the Publication It Deserves,' *Biblical Archaeology Review* 31/4 (2005): 41; Na'aman, 'When and How 24-27; Kay Prag, Review of *The Renewed Archaeological Excavations at Lachish (1973-1994),* by David Ussishkin, *Levant* 39 (2007) 249-252; Rainey, 'Stones for Bread,' 142; and Zimhoni, *Iron Age Pottery*, 200.

41 'Sheshonq,' 23.

42 *Renewed*, 83.

43 'Close Yet Apart,' 45.

44 *Biblical Lachish*, 16, 215.

45 Rainey, 'Stones for Bread,' 142.

46 *Iron Age Pottery*, 65, 171.

excavation, it was frequently difficult to differentiate between them.' Moreover, no evidence of widespread burning has been recognized. Paleographic evidence from an attempted abecedary on the palace C stairway seems to indicate that Uzziah erected podium C during this rebuilding process or slightly later during his reign.[47] Some incised drawings adjacent to the inscription are of significant interest as well.[48]

Consequently, based upon an undeniable continuity in the material culture, it may be assumed that the city was reoccupied and rebuilt shortly, if not immediately following the end of Level IV, under royal initiative as well as by the surviving Judahite inhabitants.

Discussion

While dating the demise of Lachish Level IV remains a disputed issue, the relative chronology of this event seems to have recently narrowed in light of the excavations at Tell es-Safi and Beth Shemesh. Both Ussishkin and Shai and Maeir[49] recognize close similarities between the pottery of Lachish Level IV with that from Tell es-Safi A3. Similarly, the pottery of Lachish IV and Beth Shemesh 3 is a mixture of Iron Age IIA and early Iron Age IIB types, which correspond to the late ninth-early eighth century BC.[50] Based upon these recent observations and comparisons of these two ceramic assemblages from Lachish and Tell es-Safi, the data seems to point to a destruction date close to 800 BC, as Ussishkin[51] now acknowledges. The resultant task for the historian hinges upon identifying possible correlations with recorded events dating to approximately the same time, noting that the current lack of broad exposure of this level outside of Area S severely limits our understanding of the circumstances surrounding the demise of Lachish Level IV.[52]

The present absence of evidence for a deliberate destruction, specifically the lack of a burn layer, is also intriguing and seems to correlate well with actions relating to the defeat of

47 See André Lemaire, 'Hebrew Inscriptions, Section A: Ostraca and Incised Inscriptions,' in *The Renewed Archaeological Excavations at Lachish (1973-1994)* 4, ed. David Ussishkin (Tel Aviv: Emery and Clair Yass Publications in Archaeology, 2004), 2115. Lemaire concurs with the early eighth century B.C. dating given this same inscription by David Diringer, 'Early Hebrew Inscriptions' in *Lachish III: The Iron Age (Text)* and *(Plates)*, by Olga Tufnell (London: Oxford, 1953), 357-358 and Plate 48B:3; and William Foxwell Albright (as noted in Tufnell, 'Excavations at Tell Ed-Duweir, Palestine, Directed by the Late J. L. Starkey, 1932-1938: Some Results and Reflections,' *Palestine Exploration Quarterly* 82 (1950): 77; eadem, *Lachish III*, 118; Wright, Review of *Lachish III*, *VT* 5: 100.

48 The drawings, published in Tufnell (*Lachish III*, 118 and Figure 10) and Ussishkin (*Biblical Lachish*, 219) crudely depict a roaring lion with a protruding tongue, two concentric circles as found on some *lmlk* style jars, and some additional unintelligible markings. Both the roaring lion and concentric circles are well known royal symbols of Judah and it is tempting to suggest that these drawings may actually reflect similar, but much larger and more lavish designs that adorned various walls and objects throughout the royal complex at Lachish. More importantly, the concentric circles drawn in this probable eighth century B.C. context arguably provide evidence for their use as a Judahite, not Assyrian, symbol of hegemony. Perhaps this doodling reflects the rudimentary efforts of a palace guard to duplicate the painted or carved royal symbols displayed on the palace walls above him during some idle time while stationed at his post on this lateral access stairway.

49 Ussishkin, *Biblical Lachish*, 212-15; Shai and Maeir, 'Late Iron Age IIA', 353-55.

50 Mazar and Panitz-Cohen, *Timnah*, 275-75; Faust, 'Jerusalem's Western Hill,' 107 n. 13; Fantalkin and Finkelstein, 'Sheshonq,' 22-24; Bunimovitz and Lederman, 'Close Yet Apart,' 43-45; eidem, *Tel Beth-Shemesh*, 332-38; Na'aman, 'When and How,' 24-25; eadem, 'Kingdom of Judah,' 253.

51 *Biblical Lachish*, 16, 212-15.

52 Note the observation by Bunimovitz and Lederman ('Close Yet Apart ,' 43 n. 5) regarding Zimhoni's (*Iron Age Pottery*, 69; eadem, 'Pottery of Levels V and IV,' 1652) assessment of the Level V and IV pottery, which seems to contradict the conclusions of the excavators (Ussishkin, *Renewed*, 83; Barkay and Ussishkin, 'Area S,' 445). Again, the recently completed excavations at Lachish will provide additional data for this poorly represented occupational level. For a recent discussion of the chronology of the Iron Age IIA-B transition, see Bunimovitz and Lederman (*Tel Beth-Shemesh*, 357-82).

Amaziah of Judah by Jehoash of Israel, as recorded in parallel biblical accounts. While many scholars may ridicule this link, viewing it as an overly simplistic, if not naïve correlation between biblical records and archaeological data, they must also recognize, as Dever[53] reminds us, that text-based chronologies are inevitable in historical archaeology. Sequential ceramic typologies provide a relative chronology, but absolute chronologies must come from fixed dates recorded or indicated in literary and historical sources. We must also acknowledge that some established Iron Age absolute dates derive, at least in part, from biblical historical and chronological data. Consequently, since the fiery destruction of Beth Shemesh Stratum 3 seems to fit the circumstances immediately after Amaziah's rout, the victorious Israelite soldiers most likely sacked Lachish and other, now vulnerable cities after defeating Judah's army.[54] From the data now available, the destructive activity at Lachish seems to resemble what Jehoash oversaw his army carry out at Jerusalem (2 Kgs 14:13-14; 2 Chr 25:23-34). Chr also mentions an earlier swath of destruction and violence committed by the Israelite mercenaries dismissed by Amaziah before his attack on Edom. 'But the men of the army whom Amaziah sent back, not letting them go with him to battle, fell upon the cities of Judah, from Samaria to Beth-Horon, and killed three thousand people in them, and took much spoil' (2 Chr 25:13).[55] The exclusion of Lachish and other cities in the Shephelah in this text is significant. Despite chronological correlations with their acts and circumstances to the material evidence unearthed at Lachish, its remoteness from the pillaging troops, coupled with its strong fortifications, like those of Jerusalem, provided a formidable deterrent for the marauding Israelite mercenaries. However, when news of the rout of Judah's army and capture of Amaziah reached Lachish, all attempts to organize a defense for Lachish likely evaporated. Left exposed without a garrison, Lachish probably fell or capitulated after only minimal resistance. In his review of Ussishkin's final report, Peter James[56] correctly observes that the lack of a burn layer does not rule out a

53 'Of Myths and Methods,' *Bulletin of the American Schools of Oriental Research* 277/278 (1990): 121-22; idem, 'Some Methodological Reflections on Chronology and History-Writing,' in *The Bible and Radiocarbon Dating: Archaeology, Text and Science* (eds. Thomas E. Levy and Thomas Higham; London: Equinox, 2005), 413.

54 While Jehoash's troops possibly sacked other nearby towns and cities, the archaeological evidence from Azekah, a prominent Judahite site between Beth Shemesh and Lachish, is fragmentary with inconclusive findings. The excavators claim the Iron IIA pottery recovered thus far appears to predate the Tell es Safi A3 repertoire (e.g., Lipschits; Gadot and Oeming 2017: 13-15). Firm evidence of a late ninth-early eighth century B.C. destruction is apparently also lacking at nearby Tel Burna (e.g., Shai, 'Tel Burna: A Judahite Fortified Town in the Shephelah, in *The Shephelah during the Iron Age: Recent Archaeological Studies* (eds. Oded Lipschits and Aren M. Maeir; Winona Lake: Eisenbrauns, 2017), 48-50). Tel Zayit, which competes with Tel Burna as a candidate for biblical Libnah, has a ninth century (ca. 830) B.C. destruction level that the excavator attributes to Hazael on the basis of strong parallels with the Tell es Safi A3 assemblage (Ron E. Tappy, 'The Archaeology and History of Tel Zayit: A Record of Liminal Life,' in *The Shephelah during the Iron Age: Recent Archaeological Studies* (eds. Oded Lipschits and Aren M. Maeir; Winona Lake: Eisenbrauns, 2017), 167-68). Other sites exhibiting strata with ca. 800 B.C. termination dates may include Arad XI, Beer Sheba IV and Tell Halif VIC.

55 Chr's brief report here, for which DtrH offers no parallel account, presents us with a crux. Plundering cities 'from Samaria to Beth-horon' falls almost entirely within the traditional boundaries of the kingdom of Israel, not Judah, and strangely follows a north to south-southwest trajectory, which does not reflect the expected route these returning mercenaries would follow. Opinions and proposed solutions differ among scholars (see Hugh G.M. Williamson, *1 and 2 Chronicles* (New Century Bible Commentary; Grand Rapids: Eerdmans, 1982), 328-30; R. B. Dillard, *2 Chronicles* (Word Biblical Commentary 15; Waco: Word, 1987), 200-201 and Sara Japhet, *I & II Chronicles: A Commentary* (The Old Testament Library; Louisville: Westminster/John Knox 1993), 865 for suggested readings, interpretations and resolutions). The historical veracity of this text is also debated (e.g., Ehud Ben Zvi, 'A House of Treasures: The Account of Amaziah in 2 Chronicles 25–Observations and Implications,' *Scandinavian Journal of the Old Testament* 22 (2008): 79-82; Ralph W. Klein, 'The Chronicler's Theological Rewriting of the Deuteronomistic History: Amaziah, a Test Case,' in *Raising Up a Faithful Exegete: Essays in Honor of Richard D. Nelson* (eds. K. L. Noll and Brooks Schramm. Winona Lake: Eisenbrauns, 2010), 242, 245). One may also argue that, rather than Jehoash, perhaps these disenchanted and angry soldiers of fortune actually sacked Lachish. However, the systematic destruction wrought upon the fortifications and Palace-Fort of Lachish Level IV seem to bear the marks of a concentrated, well-organized effort carried out by an army following clearly stated orders. The geographical distance of Lachish from the towns mentioned in Chr's account all but severs any link between the two events.

56 Review of *The Renewed Archaeological Excavations at Lachish (1973-1994)* by David Ussishkin, *Palestine Exploration Quarterly*

destruction by a human agent. If Jehoash indeed was responsible for Level IV's destruction, he undoubtedly singled out all defensive, royal and administrative structures as priority targets for demolition by his soldiers, which seems to align with the archaeological evidence at Lachish as well as his recorded actions against Jerusalem.

The biblical chronology for this period has long been problematic until Thiele[57] proposed a solution utilizing co-regencies and a non-accession year system in both Israel (Jehoash and Jeroboam II) and Judah (Amaziah and Azariah/Uzziah) that not only brought chronological harmony to the apparent 24-year discrepancy present in the biblical text, but also provided an exact chronology for the events surrounding Amaziah's defeat. Thiele[58] reconstructs the chronology in the following manner. Before facing Judah in battle, Jehoash installed Jeroboam II as co-regent. When Jehoash captured Amaziah at Beth Shemesh and took him as a prisoner, the people of Judah placed Azariah / Uzziah on the throne. After the death of Jehoash, Amaziah was paroled and returned to Jerusalem, where he lived for fifteen more years. While the biblical writers include all of these years in Amaziah's total reign, Azariah/Uzziah, in effect, was the sole ruler. Following Thiele's chronological system, the destructive acts of the Israelite mercenaries can be dated to the spring or summer of 793 BC. The battle at Beth Shemesh, its destruction, as well as the destruction and plundering of Jerusalem and Lachish Level IV took place during the spring-summer of 792 BC. Likewise, the installation of Azariah / Uzziah as king also occurred during late spring or summer of the same year. While the biblical evidence is not conclusive, the circumstances described and nature of the narratives strongly imply and sufficiently demonstrate the need for an immediate transfer of power in Judah in the wake of this national catastrophe. The proposed 792 BC date for these events effectively correlates the archaeological evidence, the biblical accounts and Thiele's chronology, which he first proposed without reference to or correlation with archaeological data available at the time.[59] The close resemblance of the Beth Shemesh Stratum 3 pottery to that of Lachish IV strengthens this interpretation and seems to establish an important new chronological date for the archaeology of the Iron Age II in Judah and the southern Levant.

The converging lines of evidence outlined above do not seem coincidental. Sacking such a well-fortified administrative city as Lachish Level IV was a horrendous event for Judah and certainly worthy of mention by the biblical writers save for two factors. First, Lachish probably capitulated or quickly fell when news of the debacle at Beth Shemesh reached its walls. It is unlikely that the formidable defenses of Lachish were manned by more than a handful of soldiers as most of the garrison probably left to fight at Beth Shemesh, only to meet their defeat on the field of battle. Secondly, the sack of Jerusalem itself, the extensive breaching of its wall,[60]

---

139 (2007): 214-15.

57 *Mysterious Numbers²*, 76-89; *Chronology*, 39-45; *Mysterious Numbers³*, 111-23.

58 *Chronology*, 42-43.

59 The novel suggestion put forth by Shmuel Yeivin ('The Divided Kingdom: Rehoboam-Ahaz/Jeroboam-Pekah,' in *The World History of the Jewish People. The Age of the Monarchies: Political History* 4/1 [ed. Avraham Malamat; Jerusalem: Massada,1979], 161) that Jehoash died shortly after the Beth Shemesh debacle from combat wounds does not correlate with the 782 B.C. date for his death given in Thiele's (*Chronology*, 39-45, 75; *Mysterious Numbers³*, 111-123, 217) chronology, but it does perhaps provide an explanation behind the vengeful actions of his troops at Jerusalem and, in our opinion, at Lachish.

60 The Hebrew term פרצ 'breach' translated as 'broke down' the walls (2 Kgs 14:13; 2 Chr 25:23) is used again by Chr in his description of Uzziah's actions in Philistia, specifically his breaching the walls of Gath, Yavneh and Ashdod (2 Chr 26:6). Chr's choice of verb may reflect reciprocal retributive acts carried out to restore national honor after the embarrassing debacle at Beth Shemesh and the sack of Jerusalem and other key cities. Nevertheless, Uzziah shifted his focus and waged war against Philistine, not Israelite cities since Jeroboam II, potentially a most formidable opponent, was now his ally. Jehoash's destructive dismantling of 400 cubits of Jerusalem's wall from the Ephraim Gate to the

the plundering of the Temple, the royal palace and the treasury by Jehoash far eclipsed the similar fate suffered at Lachish, especially through the lens of the biblical historians.

The recent convergence of views regarding the similar archaeological chronologies of Tell es-Safi stratum A3, Beth Shemesh Level 3, and Lachish Level IV based upon ceramic evidence argue that the termination of Lachish Level IV occurred sometime around 800 BC. The biblical chronological data marshalled earlier by Thiele posited that Jehoash defeated Amaziah at Beth Shemesh in 792 BC, resulting in Uzziah's coronation shortly later. The merging of these events and dates is highly significant and points to the likelihood that Jehoash sacked and perhaps burned parts of Lachish, but probably refrained from systematically destroying the entire city, apart from dismantling part of the city's defensives and its administrative structures.

As a follower of the low chronology espoused by his Tel Aviv University colleague Israel Finkelstein, Ussishkin undoubtedly recognizes that accepting a ca. 800 BC transition date for Lachish Levels IV-III creates a chronological dilemma for Levels V and IV. Essentially, both of these occupational strata have to be squeezed into a century (900-800 BC). Consequently, Ussishkin resolves this issue by positing a fifty-year span for each level. Despite the fact that he observed at least four distinct occupational phases of Level IV, Ussishkin based his conclusions upon rather vague 'general considerations and the chronology of parallel sites' rather than upon a trajectory of stratigraphic evidence, which seems to indicate a longer time span.[61]

Conclusion

Since the publication of Ussishkin's evidence conclusively linking the destruction of Lachish level III with Sennacherib's 701 BC campaign, scholarly consensus equates the chronological division between Iron IIB and Iron IIC with ca. 700 BC.[62] However, no such consensus exists for determining a date for the transition from Iron IIA to Iron IIB. Recent excavations at Beth Shemesh and Tell es-Safi reveal two chronologically close destruction layers. A catastrophic event dating sometime before the end of the ninth century BC at Tell es-Safi Stratum A3, and another at Beth Shemesh Level 3, which the excavators now date to ca. 790 BC.[63] Furthermore,

---

Corner Gate likely represents the northern defenses of the city, which are the most vulnerable from a topographical standpoint. Gabriel Barkay, 'An Unknown City Wall in Jerusalem of First Temple Period,' *New Studies on Jerusalem* 7 (2001): 39-44, 8* (Hebrew with English summary), convincingly argues that this description reflects that at least part of the Western Hill was already fortified in the ninth century B.C. and most certainly by 792 B.C. Barkay's position is strengthened by the evidence presented recently by Avraham Faust, 'On Jerusalem's Expansion during the Iron Age II,' in *Exploring the Narrative: Jerusalem and Jordan in the Bronze and Iron Ages* (eds. Eveline van der Steen, Jeannette Boertien, and Noor Mulder-Hymans; Library of Hebrew Bible/Old Testament Studies 583; London: Bloomsbury, 2014), 256-285. See also the important studies by Na'aman arguing that Jerusalem's Western Hill underwent a slow, evolutionary growth over two and a half centuries, 'When and How,' 24-27, 44-45; idem, 'Dismissing the Myth of a Flood of Israelite Refugees in the Late Eighth Century BCE' *Zeitschrift für die Alttestamentliche Wissenschaft* 126 (2014): 1-14. See further the recent paper by Joe Uziel and Nahshon Szanton, 'New Evidence of Jerusalem's Urban Development in the 9th Century BCE,' in *Rethinking Israel: Studies in the History and Archaeology of Ancient Israel in Honor of Israel Finkelstein* (eds. Oded Lipschits, Yuval Gadot, and Matthew J. Adams (Winona Lake: Eisenbrauns, 2017), 429-39.

61 See Ussishkin, *Biblical Lachish*, 16, 203-10. In his latest study of this issue, Ussishkin ('Gath, Lachish and Jerusalem,' 136-38) remarks that Lachish Level IV existed 'for a long period of time' and '(t)he pottery assemblage of Level III... radically differs from that of Level IV, a change that must have taken some time to crystalize.' Dever, (Review of *The Renewed Archaeological Excavations at Lachish (1973-1994)* by David Ussishkin, *Bulletin of the American Schools of Oriental Research* 340 [2005], 83-84) offers some interesting additional insights on this issue.

62 Some recent attempts to lower the transition date of Iron IIB-IIC into the seventh century B.C. have little basis in fact and remain unconvincing. In some cases, arguments for this position appear to be ideologically driven.

63 Bunimovitz and Lederman, 'Border Communities,' 136; eidem, 'Close Yet Apart,' 43-45; Shlomo Bunimovitz; Zvi Lederman, and Nicoletta Momigliano, 'Introduction,' in *The Excavations of Beth Shemesh, November—December 1912, by Duncan Mackenzie*, (ed. Shlomo Bunimovitz, Zvi Lederman, and Nicoletta Momigliano; The Palestine Exploration Fund Annual 13; London: Routledge , 2016), 9,18.

the pottery from these two levels display close similarities with Lachish level IV, prompting Ussishkin to date the end of Lachish level IV to about 800 BC as well. Recognizing the coincidental convergence with the biblical record, Beth Shemesh's excavators Bunimovitz and Lederman quite reasonably associate the city's Level 3 destruction to the battle between Jehoash of Israel and Amaziah of Judah. Israel's capture and probable detention of Amaziah (2 Kgs 14:13-17[64] promptly forced 'all the people of Judah' to make Uzziah king (2 Kgs 14:21). A rapid succession of events: a disastrous military defeat, the subsequent sack of Beth Shemesh and Jerusalem, coupled with the capture and probable extended confinement of their king; created an exceptional set of circumstances that forced in an immediate regnal change in Judah, which in turn significantly provides two converging lines of evidence from archaeological and biblical sources. Years ago, chronologist Edwin R. Thiele posited an exact chronological date of 792 BC for the defeat and capture of Amaziah and the ascension of Uzziah. The relative dating of Lachish IV and Beth Shemesh Level 3 now corresponds remarkably well to these data and indeed corroborates Thiele's chronological model. Furthermore, it appears likely that Jehoash's retaliation against Judah extended beyond Beth Shemesh and Jerusalem, involving probable destructive acts against other Judahite cities, most notably at Lachish. If our assertion is correct, we not only propose an exact date of 792 BC for the end of Lachish IV and Beth Shemesh 3, but also suggest that this date serve as a most appropriate historical benchmark for the transition from Iron IIA from Iron IIB in Judah.[65]

---

64 e.g., Thiele, *Chronology*, 41-43; Dillard, *2 Chronicles*, 198; Japhet, *I & II Chronicles*, 870; J. Maxwell Miller and John H. Hayes, *A History of Ancient Israel and Judah.* Second edition. (Louisville: Westminster/John Knox, 2006), 351-52.

65 This study was initially conceived and written in 2016, which marked the thirtieth anniversary of Edwin R. Thiele's (1895-1986) death and is presented in remembrance of this respected biblical scholar. Professor Thiele served as Professor of Antiquity at the Seventh Day Adventist Theological Seminary at Andrews University until his retirement and during his career, Thiele brilliantly harmonized what had formerly appeared to be two hopelessly conflicting sets of royal chronological records from the kingdoms of Israel and Judah. By properly understanding the co-regencies and recognizing the variant calendars utilized by these two kingdoms, Thiele first presented, then refined his conclusions, essentially unlocking their respective regnal dates, and strongly supporting, if not verifying the historical validity of these ancient records. For many scholars, Thiele's chronological system continues to be an indispensable component for constructing the history of the Israelite and Judahite monarchies and inserting them in to their proper historical and geopolitical context. His masterful historical and chronological observations play an essential role for the arguments expressed here as well. Appreciative essays and shorter notes that highlight the importance of Thiele's contributions include words of admiration and support from such luminaries as Siegfried H. Horn ('From Bishop Ussher to Edwin R. Thiele,' *Andrews University Seminary Studies* 18 (1980): 37-49), Donald J. Wiseman (*1 & 2 Kings: An Introduction and Commentary* [Tyndale Old Testament Commentary 9; Downers Grove: Inter-Varsity, 1993]: 27), and Anson F. Rainey (Review of *An Introduction to the History of Israel and Judah*, by J. A. Soggin, *Journal of the American Oriental Society* 116 [1996]: 546-48). Recent expressions of respect for Thiele's work include the honorific title chosen for a paper by D. Fouts ('The Incredible Numbers of the Hebrew Kings,' in *Giving the Sense: Understanding and Using the Old Testament Historical Texts* [eds. D. M. Howard Jr., and M. A. Grisanti; Grand Rapids: Kregel, 2003], 283-299) and an appreciative article by Rodger C. Young ('Three Verifications of Thiele's Date for the Beginning of the Divided Kingdom,' *Andrews University Seminary Studies* 45/2 [2007]: 163-89).

The Hyksos and the Exodus: Two 400-Year Stories

Peter Feinman

There are two noteworthy four-hundred year traditions from the ancient Near East. In one tradition carved in stone, Ramses II honors the legacy of the Hyksos in Egypt.[1] Ramses documents the exact date of the commemoration: year 400, month 4, season 3, day 4.[2] In the other tradition written perhaps originally on papyrus, Yahweh informs Abram that his descendants will experience oppression in Egypt for 400 years (Gen. 15:13). The date of origin for this tradition is debated.[3] The idea of a connection between the two is not new but determining the exact connection has been problematic. Manetho and Josephus in ancient times and some Egyptologists in modern times have suggested a Hyksos-Exodus link of some kind (see below).[4]

400 is not a number that appears to possess any symbolic, magic, or divine powers. It is not a measure of a natural calendric cycle. It is not a measure of an astronomical phenomenon. Nor does it factor into components that call to mind something special. What then is it? For Egypt the time ranges from the Second Intermediate Period to the New Kingdom, from the time of the Hyksos (roughly 1650-1550 BCE) to before Merneptah, the son and successor to Ramses, who mentions Israel (1206 BCE), two events not normally linked together in a single time period by academics. Egyptologists routinely separate these two time periods thereby obscuring the possibility that the Egyptians (Ramses II) may have thought of them as a single period. For Israel, based on the traditional view of Ramses II as the Pharaoh of the Exodus, the range is the same from the 17th to 13th centuries BCE. How is it that these two cultures employed the same unusual number to refer to the same time period?

This investigation to resolve the question starts at the very mathematical beginning, a very good place to start and takes it one factor at a time. In the beginning was the well-known number four. There are the four compass directions, front back, left, and right. There are

1 For the Hyksos, see Manfred Bietak, 'The Aftermath of the Hyksos in Avaris,' in *Culture Contacts and the Making of Cultures: Papers in Homage to Itamar Even-Zohar* (ed. Rakefet Sela-Sheffy and Gideon Toury; Tel Aviv: Unit of Culture Research Tel Aviv University, 2011), 19-65; Manfred Bietak, 'On the Historicity of the Exodus: What Egyptology Today Can Contribute to Assessing the Sojourn in Egypt,' in *Israel's Exodus in Transdisciplinary Perspective: Text, Archaeology, Culture, and Geoscience* (ed. Thomas E. Levy, Thomas Schneider and William H.C. Propp; Heidelberg-New York- Dordrecht-London: Springer, 2015), 17-36; Marcel Marée (ed.), *The Second Intermediate Period (Thirteenth-Seventeenth Dynasties): Current Research, Future Prospects* (Orientalia Lovaniensia Analecta 192; Leuven: Peters, 2010); Eliezer D. Oren (ed.), *The Hyksos: New Historical and Archaeological Perspectives* (University Museum Symposium Series 8; Philadelphia: University Museum, University of Pennsylvania, 1997); Donald B. Redford, 'The Hyksos Invasion in History and Tradition, *Orientalia* NS 39 (1970): 1-51; Donald B. Redford, and James Weinstein, 'Hyksos,' *Anchor Bible Dictionary* III: 341-348; K. S. B. Ryholt, *The Political Situation in Egypt During the Second Intermediate Period c. 1800-1550 B.C.* (Copenhagen: K.S.B. Ryholt and Museum Tusculanum Press, 1997); John Van Seters, *The Hyksos: A New Investigation* (New Haven: Yale University Press, 1966); Anna-Latifa Mourad, *Rise of the Hyksos: Egypt and the Levant from the Middle Kingdon to the Early Second Intermediate Period* (Oxford, Archaeopress Publishing, 2015).

2 For the 400-year stela, see Hans Goedicke, 'Some Remarks on the 400-Year Stela,' *Chronique d'Egypte* 41 (1966): 23-37; Hans Goedicke, 'The '400-Year Stela' Reconsidered,' *Bulletin of the Egyptological Seminar* 3 (1981): 25-42; Labib Habachi, 'The Four Hundred Year Stela Originally Standing in Khatâ'na-Qantîr or Avaris-Piramesse,' in *Actes du XXIXe Congrès des Orientalistes Section Égyptologie Vol. 1* (ed. Goerges Posener; Paris: L'Asiatheque, 1975), 41-44; M. Pierre Montet, 'La Stèle de l'An 400 Retrouvée,' *Kemi* 4 (1933): 191-215; Jean Yoyotte, 'Les Stèles de Ramses II à Tanis: Première Partie' *Kemi* 10 (1949): 58-74.

3 Readers interested in pursuing this debate are invited to start with the Genesis commentaries.

4 For Josephus on Mantheo's portrayal of the Hyksos/Jews in Egypt see *Contra Apionem* I:72-105 and 227-288.

the four corners of the universe. Despite the general awareness that the earth is not flat, the reference to the four corners for the earth persists because of its imagery. As a metaphor it still works. The continued use of the two-dimensional (Mercator) maps in books, hung on walls, and shown on screens, reinforces the validity of the concept of a four-cornered earth.

In ancient times before such information about the earth as a globe was known, the image of a flat-earth with four corners was even more entrenched. The concept was fully integrated into the Mesopotamian cultures where references to the four quarters of the universe were common. Rulers seeking to brag about the extent of their domain were prone to dispense with precise geographical details in favor of the bombastic claim to rule the entire world from the Lower Sea (Persian/Arabian Gulf) where the sun rose to the Upper Sea (Mediterranean) where the sun set. Piotr Michalowski observes:

> Not to be content to be kings of Sumer and Akkad, these [Akkadian] rulers added still another forceful epithet, 'king of the four corners of the universe,' or, in Sumerian, 'kings of the heaven's four corners,' in a sense driving home the notion of 'everything.'[5]

This sense of 'everything' through the use of 'four' continued across the millennia in Mesopotamian times from Akkadians to Assyrians.

> I (Ashurnasirpal II), great king, king of the universe, king of Assyria, king of all the four quarters
>
> I, Ashurnasirpal (II) ...shepherd of the four quarters, who has brought all peoples under one authority, avenger of Assyria, who extends the borders of his land... [6]

In the 7th-century BCE, Ashurbanipal channeled the sentiments of his 9th-century predecessor:

> At the proclamation of my honored name [upon installation as king], the four regions of the world were glad and rejoiced. The kings of the Upper and Lower seas, vassals subject to my father, sent me tidings of their joy at my installation of the kingly office....The lands were quiet. The four regions of the world were in perfect order, like the finest oil.[7]

The sense from these usages is one of totality, of completeness, of fulfillment. The number four representing our ability to see forward, behind, to left, and to the right encompasses everything ... excluding up (the heavens) and down (land of the dead)!

Four certainly is known in the biblical tradition and in the same cosmic sense. There are the four rivers which water the entire world in the garden story:

> *A river flowed out of Eden to water the garden, and there it divided and became four rivers* (Gen. 2:10).

5 Piotr Michalowski, 'Masters of the Four Corners of the Heavens: Views of the Universe in Early Mesopotamian Writings,' in *Geography and Ethnography: Perceptions of the World in Pre-modern Societies* (eds. Kurt A. Raaflaub and Richard J.A. Talbert; Malden, MA: Wiley-Blackwell, 2010), 147-168, here 153.

6 Albert Kirk Grayson, *Assyrian Royal Inscriptions 2* (Records of the Ancient Near East; Wiesbaden: O. Harrassowitz, 1976), §§ 712, 714.

7 Daniel David Luckenbill, *Ancient Records of Assyria and Babylonia* (Chicago: University of Chicago Press, 1927), #987.

There are the four cities over which Nimrod rules the entire world if one ignores Egypt:

> *The beginning of his kingdom was Babel, Erech, Accad, and Calneh in the land of Shinar* (Gen. 10:10).

Most likely these four names also suggest a chronological unity in addition to the geographical one especially if Calneh was substituted for Calah in Neo-Assyrian times. The names then reflect the four great empires of the ancient Near East: the Sumerian of Gilgamesh, the Akkadian of Sargon the Great, the Babylonian of Hammurabi and/or Nebuchadnezzar I, and the Assyrian of Tukulti-ninurta. As a result, in addition to a geographical unity and totality there is a temporal one as well.[8]

This sense of four kingdoms over time representing a single and complete sequence persisted in ancient Near East and biblical traditions.[9]

> *After you shall arise another kingdom inferior to you, and yet a third kingdom of bronze, which shall rule over all the earth. And there shall be a fourth kingdom, strong as iron, because iron breaks to pieces and shatters all things; and like iron which crushes, it shall break and crush all these* (Daniel 2:39-40).

> *Daniel said, 'I saw in my vision by night, and behold, the four winds of heaven were stirring up the great sea. And four great beasts came up out of the sea, different from one another. The first was like a lion and had eagles' wings. Then as I looked its wings were plucked off, and it was lifted up from the ground and made to stand upon two feet like a man; and the mind of a man was given to it. And behold, another beast, a second one, like a bear. It was raised up on one side; it had three ribs in its mouth between its teeth; and it was told, `Arise, devour much flesh.' After this I looked, and lo, another, like a leopard, with four wings of a bird on its back; and the beast had four heads; and dominion was given to it. After this I saw in the night visions, and behold, a fourth beast, terrible and dreadful and exceedingly strong; and it had great iron teeth; it devoured and broke in pieces, and stamped the residue with its feet. It was different from all the beasts that were before it; and it had ten horns* (Daniel 7:2-7).

The exact identity of the four kingdoms intended by Daniel is something of a parlor game among academics. As empires rise and fall, it becomes necessary to accommodate the changes to the fixed four-kingdom format in universal history with the tantalizing question of what and when would be the fifth kingdom.[10]

---

8 For Calah and the cities of Nimrod, see Peter Feinman, *Jerusalem Throne Games: The Battle of Bible Stories after the Death of David* (Oxford: Oxbow Books, 2017), 321-324, 379.

9 See Cynthia Edenburg, 'David, the Great King. King of the Four Quarters: Structure and Significance in the Catalog of David's Conquests (2 Samuel 8:1-14, I Chronicles 18:1-13),' in, *Raising Up a Faithful Exegete: Essays in Honor of Richard D. Nelson* (eds. K. L. Noll and Brooks Schramm; Winona Lake: Eisenbrauns, 2010), 159-175.

10 For Daniel and the four kingdoms tradition in the ancient Near East, see John J. Collins, 'The Mythology of Holy War in Daniel and the Qumran War Scroll: A Point of Transition in Jewish Apocalyptic,' *Vetus Testamentum* 25 (1975):596-612, here 601-602; John J. Collins, *The Apocalyptic Vision of the Book of Daniel* (Harvard Semitic Monographs 16; Missoula, MT: Scholars Press, 1977), 36-43; David Flusser, 'The Four Empires in the Fourth Sybil and in the Book Daniel,' *Israel Oriental Studies* 2 (1972):148-175, reprinted in David Flusser, *Judaism and the Origins of Christianity* (Jerusalem: The Magness Press, 1988), 317-344; Gerhard Hasel, 'The Four World Empires of Daniel 2 Against Its Near Eastern Environment,' *Journal for the Study of the Old Testament* 12 (1979):17-30; Ernest C. Lucas, 'The Origin of Daniel's Four Empires Scheme re-Examined,' *Tyndale Bulletin* 40 (1989):185-202; Arnaldo Momigliano, 'The Origins of Universal History,' in *The Poet and History: Essays in Literacy and Historical Biblical Criticism* (ed. Richard Friedman; Chico, CA: Scholars Press, 1983), 133-148; H. H. Rowley, *Darius the Mede and the Four World Empires in the Book of Daniel - A Historical Study of Contemporary Theories* (Cardiff: University of Wales Press Board, 1935); Joseph Ward Swain, 'The Theory of the Four Monarchies–Opposition History under the Roman Empire,' *Classical Philology* 35 (1940):1-21.

Finally, there are the four kings of chaos whom Abram combats.

> *In the days of Amraphel king of Shinar, Arioch king of Ellasar, Chedorlaomer king of Elam, and Tidal king of Goiim...* (Gen. 14:1).

Just as kings like to portray themselves as rulers of the four quarters of the universe, so too, they claimed to be protecting the civilized world from the forces of chaos on each side. When biblical archaeology accelerated during the waning decades of the 19th century, there was a concerted effort to locate these kings of the east from the biblical record. If one could prove that Amraphel was Hammurabi, for example, then one would have demonstrated an incontrovertible link between the biblical account and secular history.[11] The quest for physical literalness undermined the metaphorical message delivered by the story. The Elamites were the paramount force of chaos for the Mesopotamians for their sack of Babylon, somewhat like the Huns later became due to their wars against Rome.[12] This sense of the four kings being an ancient version of the four horsemen of the apocalypse, a united force of chaos, has been lost. The geographical setting of the story to the literal-minded undercuts this message: the whole world is against us and it is only the warrior-shepherd of Hebron who can and does save us. In this regard the Hollywood storytellers of cosmic superhero showdowns have a better intuitive sense of the way in which ancient stories and images resonated with the audiences than traditional scholarship. The biblical use of four in these examples is meant to convey the same sense of totality and completeness as in the previous non-biblical examples.

Raising the number four by a factor of ten continues the metaphorical not literal dimension of numbers. Forty also is number well-known from the biblical tradition as seen in these examples.

> It rains for forty days and nights in the Flood story (Gen. 7:4, 12, and 17; 8:6).
>
> Isaac marries Rebekah and Esau marries Judith at age forty (Gen. 25:20; 26:34).
>
> Jacob's embalming requires forty days (Gen. 50:3).
>
> Moses's life is divided into forty years periods (Acts 7:23, 30, 36, 42).
>
> Israel wanders in the wilderness for forty years (Ex. 16:35; Num. 14:33-34; 32:13; Deut. 2:7; 8:2, 4; 29:5; Josh. 5:6; Neh. 9:21; Ps. 95;10; Amos 5:25; Acts 13:18; Heb. 3:9, 17).
>
> Moses and Elijah were on the mountain for forty days and forty nights (Ex.24:18; 34:28; Deut. 9:9, 11, 18; 10:10; I Kings 19:8).

---

11 For some early explorations of Gen. 14, see Stanley E. Hardwick, *Change and Constancy in William Foxwell Albright's Treatment of Early Old Testament History and Religion, 1918-1958* (PhD. Dissertation, New York University, 1958), 130-190; Theophilus G. Pinches, 'Certain Inscriptions and Records Referring to Babylonia and Elam and Their Rulers, and Other Matters,' *Journal of the Transactions of the Victoria Institute* 29 (1897): 43-90; Theophilus G. Pinches, *The Old Testament: In Light of the Historical Records and Legends of Assyria and Babylonia* (London: Society for Promoting Christian Knowledge, 1902), Chapter VI, Abraham - A *Short* Account of this period, with the story of Amraphel, Arioch, and Tidal, (emphasis added), 192-241; Archibald H. Sayce, 'The Fourteenth Chapter of Genesis,' *Expository Times* 4 (1892):14-19; idem, *The 'Higher Criticism' and the Verdict of the Monuments* (London: Society for Promoting Christian Knowledge, 1894), 160-171; idem, *Patriarchal Palestine* (London: Society for Promoting Christian Knowledge, 1895), 65-80; idem, 'Archaeological Commentary on Genesis [XIV],' *Expository Times* 8 (1897): 461-465; idem, 'The Archaeology of Genesis XIV,' *Expository Times* 17 (1906):498-504; idem, 'The Chedor-laomer Tablets,' *Proceedings of the Society of Biblical Archaeology* 28 (1906): 193-200 and 247-251; idem, 'The Chedor-laomer Tablets,' *Proceedings of the Society of Biblical Archaeology* 29 (1907): 7-17.

12 The famed Code of Hammurabi, emblematic of Mesopotamian civilization was found not in Babylon where Hammurabi was king but in Susa where it had been taken. A captured statue of Marduk was famously restored from the Elamites (John A. Brinkman, 'Nebukadnezar I,' *Reallexikon der Assyriologie* 9 [1999]: 191-92) just as David restored the ark from the Philistines in the biblical tradition.

The spies spy for forty days (Num. 13:25).

Caleb begins his spy mission at age forty (Josh. 14:7).

The land has rest for forty years with multiple judges and one priest (Judg. 3:11; 5:31; 8:28; I Sam. 4:18).

The Philistines rule the land for forty years and spend forty days confronting Saul (Judg. 13:1; I Sam. 17:16).

Saul, David, Solomon, and Jehoash reign for forty years (II Sam. 5:4; I Kings 2:11; 11:42; II Kings 12:1; I Chr. 29:27; II Chr. 9:30; 24:1; Acts 13:21).

Ishbaal begins his reign over Israel at age forty (II Sam. 2:10).

Hazael brings forty camel loads to his meeting with Elisha (II Kings 8:9).

Ezekiel prophesizes forty years of desolation for Egypt (29:11-13).

Jonah prophesizes the destruction of Nineveh in forty days (3:4).

The usages continue on to the New Testament as well as indicated in some examples above. In addition there are:

Jesus fasts in the wilderness forty days (Mark 1:13; Matt. 4:2; Luke 4:20).

Jesus appears to the apostles for forty days (Acts 1:3).

The conspiracy against Paul included 'more than forty' (Acts 23:13, 21).

This extensive list of the use of the number 40 across a wide range of times, people, and circumstances suggest some intrinsic value was associated with the number 40 beyond a literal meaning. When used with years, the number signifies a distinctive period of time: times of oppression, times of peace, times of wilderness wandering, the age at the initiation of an action. When employed with days, the range of activities varies. In some cases, the action is truly cosmic: the duration of the Flood, the time on the mountain of the Lord, the time of fasting by the Son of God who resists Satan. Other activities are less dramatic but still retain a sense of importance. My sense of the usage is that 4 x 10 also implies a totality, the completion or fulfillment of a measure of time, a way of marking periods or cycles, and is not to be taken literally. It signifies the right amount in time or for an action. God forbid Hazael should have brought 41 camel loads or Moses and Elijah should have remained on the mountain top for 39 days and nights. Those actions would have disrupted the cosmic order. The audience expected 40.

The number 40 also is attested outside the biblical narrative. In the Mesha Stele, Mesha, the king of Moab, declares that Israel had ruled over the land of Moab for forty years.

> Omri had taken possession of the land of Medeba, and dwelled there his days and much of his son's days, forty years.[13]

13 Bill T. Arnold and Bryan E. Beyer, *Readings from the Ancient Near East: Primary Sources for Old Testament Study* (Grand Rapids: Baker Academic, 2002), 161-162.

The more challenging question is to determine how it came to be that Mesha used the same number used so frequently in biblical accounts. In this regard, the task is similar to that between the two usages of 400 by Ramses and the story of oppression in Egypt under Ramses. Was Mesha drawing on a shared West Semitic tradition in some way that had existed from time immemorial? If so, it would prove exceedingly difficult to determine the origin of the tradition. Would it have persisted through oral tradition or was it maintained in a more formal process through scribal education?[14]

The latter option raises two possible ways the Moabites might have learned the tradition of 40 from Israel. As Mesha admits, Moab had been ruled by Israel. The Moabite alphabet essentially is the Hebrew alphabet. Had the Moabite scribes using the letters of the Hebrew alphabet also read Israelite narratives and become imbued with Israelite motifs? Did the Moabite scribes learn the Hebrew alphabet from the diplomatic exchanges through Samaria? Perhaps the Moabite scribes learn the Hebrew alphabet from the Israelites at Mount Nebo, the sanctuary to Yahweh Mesha would destroy? At this point it is impossible to know for sure.[15]

The investigation now proceeds to the next level: raising the four number by another factor of 10: 4 x 10 x 10 to 400. In this instance there are examples of precisely 400 while others seemed to be based on the tradition of 400 with some additions.

Arpachshad lives after the birth of Shelah four hundred and three years (Gen. 11:13).

Shelah lives after the birth of Eber four hundred and three years (Gen. 11:15).

Eber lives after the birth of Peleg four hundred and thirty years (Gen. 11:17).

Abraham purchases a burial plot for 400 shekels (Gen. 23:15-16).

Esau faces Jacob with 400 men (Gen. 32:6 and 33:1).

The number of virgins found in Jabesh-Gilead as wives for Benjamin was 400 (Judges 21:12).

David in a variety of incidents has 400 military men accompanying him (I Sam. 22:2; 25:13; I Sam. 30:10, 17).

14 Ironically, the lengthiest example of the use of the Hebrew alphabet from the Iron Age discovered archaeologically may be the Mesha Stele, see Heinz Bernhardt, 'The Political Situation in the East of Jordan during the Time of King Mesha,' in *Studies in the History and Archaeology of Jordan, Vol. I* (ed. Adnan Hadido; Amman: Department of Antiquities, 1982), 163-167; M. Clermont-Ganneau, 'La Stèle de Mésa: Examen Critique du Texte,' *Journal asiatique* 9 (1887):72-112; Andrew Dearman, ed., *Studies in the Mesha Inscription and Moab* (Archaeology and Biblical Studies 2; Atlanta: Scholars Press, 1989); John A. Emerton, 'The Value of the Moabite Stone as an Historical Source,' *Vetus Testamentum* 52 (2002): 483-492; André Lemaire, 'La Stèle de Mésha et L'Histoire de L'Ancien Israël,' in *Storia e Tradizioni di Israele: Seritti in onore di J. Alberto Soggin* (eds. Daniele Garrone and Felice Israel; Brescia: Paideia, 1991), 143-169; Nadav Na'aman, 'King Mesha and the Foundation of the Moabite Monarchy,' *Israel Exploration Journal* 47 (1997):83-92; Nadav Na'aman, 'Royal Inscription versus Prophetic Story: Mesha's Rebellion according to Biblical and Moabite Historiography,' in *Ahab Agonistes: The Rise and Fall of the Omri Dynasty* (ed. Lester L. Grabbe; Library of Hebrew Bible/Old Testament Studies 421; London: T&T Clark, 2005), 145-183, here 173; Bruce Routledge, 'The Politics of Mesha: Segmented Identities and State Formation in Iron Age Moab,' *Journal of the Economic and Social History of the Orient* 43 (2000):221-256; Bruce Routledge, *Moab in the Iron Age: Hegemony, Polity, Archaeology* (Philadelphia: University of Pennsylvania Press, 2004), 27-40, 114-153, 213-221; Seth Sanders, 'What Was the Alphabet For? The Rise of Written Vernaculars and the Making of Israelite National Literature,' *Maarav* 11 (2004):25-56, here 55; Brian B. Schmidt, 'Memorializing Conflict: Toward an Iron Age 'Shadow' History of Israel's Earliest Literature,' in *Contextualizing Israel's Sacred Writings: Ancient Literacy, Orality, and Literary Production* (ed. Brian B. Schmidt; Ancient Israel and its Literature 22; Atlanta: SBL Press, 2015), 103-132, here 121-124.

15 For the connection between Mount Nebo and Nabû, the Mesopotamian writing deity, see William M. Schniedewind, *How the Bible Became a Book* (New York: Cambridge University Press, 2004), 26.

Asherah has 400 prophets in a confrontation with Elijah while Baal has 450 (I Kings 18:19, 22).

Yahweh has 400 prophets whom Jehoshaphat calls upon (I Kings 22:6).

One suspects that the modifications of the standard 400 by an addition of 3 or 30 also were deliberate. These compound numbers combined two sets of metaphorical numbers: one based on the number 4 and one based on the number 3. The biblical usage of 3, 30, and 300 is outside the scope of this study except to note that it was highly likely that those numbers delivered metaphorical messages as well.[16]

Biblical writers did not stop with 400. They raised the number by another factor of 10 and then did so again albeit not frequently.

The Philistines kill 4000 Israelites in a battle (I Sam. 4:2).

Solomon has 4000 stalls for his chariot force (II Chr. 9:25).

Ezekiel seems partial to 4500 cubits (48:16, 30, 32-34).

Joshua has an armed force of 40,000 as does the tribe of Asher (Joshua 4:13; I Chr. 12:36).

Deborah's military age population seems to be 40,000 (Judges 5:8).

Some tribes have an (armed) population of 40,500 (Num. 1:33; 2:19; 26:18).

David slays 40,000 Aramaeans (II Sam. 10:18).

One notes the frequent military basis for these usages. The numbers relate to direct military combat or the number of people eligible for military service. This use of thousands or 'eleph' may derive from the meaning of a troop unit.[17] Nonetheless the sense of completeness remains. One may contrast the modern number of platoons to the number of platoons at full-strength. Undoubtedly a clan or tribe contributing its full complement or a king leading an army at full-strength was in issue in ancient times as well. The large numbers may convey the sense of 'all the people' in a military context.

This review of the number 4 is not meant to belittle or demean the advocates of 3, 30, and 300. In general though, I claim that the use of the numbers should not be taken literally. In this sense, numbers do have verbal values. I contend that the use of 4 and its higher factored variants base 10 are meant to convey non-numerical meanings of completion, perfection, or stages depending on the specific use.

Ramses uses 400 years in the appropriately named 'Four-Hundred Year Stela.'[18] He also uses day 4 and the month 4. James Hoffmeier characterizes this as 'odd, raising the possibility of some

16 Shemaryahu Talmon suggests there is a connection between 3 and 4. He refers to the 'Topped Triad' formula of '3 + 1' producing four as the basis for the preponderance of units of four ("400 Jahre' oder 'vier Generationen' [Gen 15,13-15]: Geschichtliche Zeitangaben oder literarische Motive?' in *Die Hebräische Bibel und ihre zweifache Nachgeschichte: Festschrift für Rolf Rendtorff zum 65. Geburtstag* (eds. Erhard Blum, Christian Macholz, and Ekkehard W. Stegemann; Neukirchen-Vluyn: Neukirchener, 1990), 13-25, here 19.

17 George Mendenhall, 'The Census Lists of Numbers 1 and 26,' *Journal of Biblical Literature* 77 (1958): 52-66.

18 Manfred Bietak suggests Ramses may have replaced a commemoration stela of the actions of grandfather Ramses I and father Seti I from the time of Horemheb ('On the Historicity of the Exodus,' 31-32).

sort of symbolism.'[19] The stela commemorates the action of his father Seti I in fusing the Baal-Seth identity in the new Egyptian capital at Avaris in the 19th Dynasty. In a sense, the action officially demarcates the cessation of the Amarna Era (chaos) and the primacy of the Baal-Seth deity at Avaris over the Amun-Re deity at Thebes in the 18th Dynasty. All these machinations automatically have political overtones. While the politics of the birth of the 19th Dynasty is beyond the scope of this article, one should remain cognizant that those developments form the backdrop to the 400-Year Stele.

Again my sense is this higher factor of 4 and 10^2 signifies a unit of completion or perfection. In this case, Ramses is referring to a period of time or cycle that presumably has now concluded.[20] As with the Mesha Stele and its use of 40, the question then becomes what was the basis for Ramses choosing to use this number in reference to the Hyksos? Did the Egyptian deliberately choose to use a West Semitic number of perfection for the completed cycle when he honored the West Semitic Hyksos or is it simply a coincidence? Given the presence of Baal and Anat in Egypt, one may reasonably presume a pharaonic familiarity with Canaanite traditions. It seems likely that Egypt first used 400 years for a Hyksos sojourn in Egypt and Israel then applied the same number for the same time period for its sojourn in Egypt.

Baruch Halpern suggests that if the Israelites scribes knew the 400-year stela, that such knowledge is evidence of the portrayal of Israel as Hyksos and the identification of Ramses as the Pharaoh of the Exodus. He asserts the Israelites linked themselves to the memory of the Hyksos in Egypt probably during the time of Solomon when relationships between the two countries were good and monuments were being relocated from Goshen/Avaris to Tanis where the 400-year stele ultimately was found. He does not appear to consider the possibility that the some Hyksos actually led the people who left Egypt in the time of Ramses II and that therefore these linkages were always part of the Israelite cultural heritage.[21]

Jan Assmann similarly notes the connection between the two 400-year traditions. First, he calls attention to the often-overlooked uniqueness of the Egyptian stela.

> It represents the first—and for a long time remained the only—instance of a historical anniversary recorded in the annals of history.

Next, he asks the question if the two 400-year traditions in Egypt and Israel are coincidental. He calls this non-coincidence a 'resonance phenomenon.' He suggests that Ramses' celebration of the establishment of the Baal-related cult of Seth at Avaris and when the Hyksos took power are part of a single action. He considers it improbable that this anniversary celebration did not extend into the Egyptian-ruled land of Canaan where Baal also was prevalent. Assmann

19 James K. Hoffmeier, 'What Is the Biblical Date for the Exodus? A Response to Bryant Wood,' *Journal of the Evangelical Theological Society* 50 (2007): 225-247, here 238 n.74.

20 A similar suggestion has been made involving the 480 years between the departure from Egypt and the commencement of the building of the Solomonic temple (I Kings 6:1) and the sojourn in Egypt for 430 years (Ex 12:40-41). While there are many genealogical accounts in the Bible there are few epochal references and 400 years is a factor in these two (see Yong Ho Jeon, 'The Retroactive Re-Evaluation Technique with Pharaoh's Daughter and the Nature of Solomon's Corruption in I Kings 1-12,' *Tyndale Bulletin* 62 (2011): 15-40, here 30.

21 Baruch Halpern, The Exodus from Egypt: Myth or Reality,' in Hershel Shanks, William G. Dever, Baruch Halpern, P. Kyle McCarter, eds. *The Rise of Ancient Israel: Symposium at the Smithsonian Institution, October 26, 1991* (Washington, DC: Biblical Archaeological Society, 1992), 86-117, here 98-101; and Baruch Halpern, 'Fracturing the Exodus, as Told by Edward Everett Horton,' in Thomas E. Levy, Thomas Schneider, and William H. C. Propp, ed. *Israel's Exodus in Transdisciplinary Perspective: Text, Archaeology, Culture, and Geoscience* (New York: Springer, 2015), 293-304, here 299.

overlooks one other possible development. Israel, being aware of the celebrated connection among Baal, Seth, and 400, rejected the message Ramses was delivering. The 400 years was a time of oppression and not celebration. The people who became Israel would base their religious identity on a deity other than Baal-Seth, one who was in opposition to Egypt, one who surpassed Baal.[22]

I propose that there is a connection between the 400 year stories and it originated in the aftermath of the Battle of Kadesh during the reign of Ramses II.[23] For Ramses the use of 400 years served a double purpose. If Ramses had wanted to use a more precise number to mark the period of the Hyksos in Egypt, he could have done so. Seti's temple at Abydos contains a 'Hall of Records' or 'Gallery of Lists' with a series of cartouches of the kings of Egypt from the dawn of time until Seti's present (excluding Amarna).[24] These cartouches create a form of timeline or chronology for the kings. The Turin King List from the Ramses's reign similarly would have enabled him to calculate a more precise number had he wanted to.[25] The issue here is not the validity of those Egyptian chronological reconstructions but the opportunity for Ramses to draw on the data had he so desired. He did not avail himself of that opportunity. He was not a grad student seeking to document a chronology, he was a political leader attempting to deliver a relevant political message.

Ramses's father had marked the dawn of a new era in Egyptian history. Seti was quite conscious of having started a new dynasty. He bore the title 'Repeater of Births' marking the beginning of a new era.[26] He explicitly characterized himself in the same terms Middle Kingdom pharaoh Amenemhet I did to launch the 12th Dynasty.[27] Seti's actions suggest a conceptual awareness of at least three periods in Egyptian history: the Old Kingdom which had collapsed thus necessitating the birth of the Middle Kingdom which then had collapsed (with Amarna?) thus necessitating the birth of a New Kingdom with the 19th Dynasty. Seti also was from a Delta-family of chariot background who made the Hyksos capital his capital.[28] Whether or not there

22 Jan Assmann, *The Invention of Religion: Faith and Covenant in the Book of Exodus* (Princeton: Princeton University Press, 2018), 36. This author will further explore these points in two forthcoming publications, *The Exodus: An Egyptian Story* (Oxford: Oxbow Press) and 'A Hyksos Levite Led Exodus in the Time of Ramses II,' Mark D. Janzen, ed., *Five Views on the Exodus* (Grand Rapids, Zondervan).

23 For the battle of Kadesh/Qadesh, see Jan Assmann, *The Mind of Egypt: History and Meaning in the Time of Pharaoh* (Cambridge: Harvard University Press, 2002), 255-271; James Henry Breasted, *The Battle of Kedesh* (Chicago: University of Chicago Press, 1903); Clive Broadhurst, 'Religious Considerations at Qadesh, and the Consequences for the Artistic Depiction of the Battle,' in *Studies in Pharaonic Religion and Society in Honour of J. Gwyn Griffiths* (ed. Alan B. Lloyd; London: The Egypt Exploration Society, 1992), 77-81; M. J. de Bruyn, 'The Battle of Qadesh: Some Reconsiderations,' in *To the Euphrates and Beyond: Archaeological Studies in Honour of Maurits N. Van Loon* (eds. Odette M. C. Haex, Hans H. Curvers, and P. M. M. G. Akkermans; Rotterdam: A. A. Balkema, 1989), 135-165; Raymond O. Faulkner, 'The Battle of Kadesh,' *Mitteilungen des Deutschen Archäologischen Instituts Abteilung Kairo* 16 (1958): 93-111; Hans Goedicke, *Perspectives on the Battle of Kadesh* (Baltimore: Halgo, 1985); Kenneth A. Kitchen, *Pharaoh Triumphant: The Life and Times of Ramesses II* (Warminster, Wilts.: Aris & Phillips, 1982), 53-61; Anthony Spalinger, *The Transformation of an Ancient Egyptian Narrative: P. Sallier III and the Battle of Kadesh* (Wiesbaden: Harrassowitz, 2002); Anthony Spalinger, 'Epigraphs in the Battle of Kadesh Reliefs,' *Eretz-Israel* 27 (2003): 222-239.

24 Peter A. Clayton, *Chronicle of the Pharaohs: The Reign-by-Reign Record of the Rulers and Dynasties of Ancient Egypt* (London: Thames & Hudson, 1994), 143.

25 For the use of the Turin Canon to reconstruct the Egyptian royal king list, see Donald B. Redford, *Pharaonic King-Lists, Annals and Day-Books: A Contribution to the Study of the Egyptian Sense of History* (Mississauga, Ont.: Benben Publications, 1986), especially 2-18.

26 Clayton, *Chronicle of the Pharaohs*, 141-142.

27 Clayton, *Chronicle of the Pharaohs*, 79.

28 R. O. Faulkner, 'Egypt: From the Inception of the Nineteenth Dynasty to the Death of Ramesses III,' in *The Cambridge Ancient History 2/1* (ed. I. E. S. Edwards, et al.; Cambridge: Cambridge University Press, 1975), 217-251, here 221.

had been any intermarriage between the Hyksos and his family is suspected but not proven and perhaps not provable.

Even if no such intermingling had occurred, I propose that in the Four-Hundred Year Stela, Ramses sought to merge the two traditions as his father had. The time of the onset of the new Egyptian dynasty was the time of the completion of a period in history. He integrated the Hyksos timeline into the Egyptian one. Instead of the Hyksos ruling during an 'intermediate period' as in Egyptology, the Hyksos were the beginning of a cycle which concluded with the post-Amarna restoration. What had been separate now became one. Baal began both periods in history. From this point forward, the two peoples were chronologically merged into a single timeline in Egyptian history. It was morning in Egypt. Here comes the sun on a new day in Egyptian history. Ramses had delivered a political message in his present through the metaphorical values of the numbers he chose to publically proclaim in the organization of temporal epochs.

Ramses, drawing on his father, had proposed a bold move. He was thinking outside the box or outside the valley and sun that traditionally defined Egyptian chronology from days to years to Sothic cycles. Egypt was the land of Pharaoh-smites-the-enemy, where people outside the land were not real humans like the Egyptians. Now the father of a daughter named after the Canaanite goddess Anat was seeking a merger of two peoples. Baal was Set and Set was Baal. Egypt and the Hyksos were part of a single shared tradition. They lived in the same city, a capital to both the Hyksos and the 19th Dynasty. Their calendars were in synch: 400 years and the start of a new dynasty.

Egypt was adjusting to the expanded arena extending beyond the river valley. Ramses sought to integrate Canaanite cultural values into the Egyptian culture, at least of its Hyksos warriors, the people among whom his own family had lived for generations. A Theban king would never do that. The one foray into a new cultural paradigm during the 18th Dynasty vanished with the death of its proponent. Now the Delta-born Ramses proposed a new way of coping with the international arena based on the reality of the Delta life - a merger of two warrior peoples to create an even stronger one that would dominate Egyptian-ruled lands. Why would an Egyptian ruler even consider such a grandiose and cultural-altering proposition?

Before answering that question, let's skip ahead to Ramses's son and successor who also dealt with Hyksos. With Merneptah, the situation takes a 180 degree turn back to the old days. There is no honoring of the Hyksos. *The Quarrel Story* paints a substantially different picture.[29] Merneptah repudiated the outreach to the Hyksos extended by Ramses. Contrary to the ministrations of Egyptologists, Seqenenre is not the hero of the original Quarrel story. He is the loser. The hero is or are the successor Pharaohs Kamose and Ahmose. Maguid Nafouz, the Egyptian Nobel-prize winner, is a more perceptive storyteller then Egyptologists, as he demonstrated in *Thebes at War*. Nafouz recognized that the story of the conflict between Egypt and the Hyksos extended beyond a single encounter between Seqenenre and Apophis. An individual monument by a king would not tell that story; a storyteller in *The Quarrel Story* could. Once one realizes that the original story continued and concluded with the success of Seqenenre's successors, as Nafouz did in his book, one gains insight into Mernenptah's Hyksos policy.[30]

29 For a recent analysis of The Quarrel Story, see Peter Feinman, 'The Quarrel Story: Egypt, the Hyksos, and Canaan,' *Conversations with the Biblical World* 35 (2015): 94-127.

30 For a historical reconstruction in a novel format, see Maguid Nafouz, *Thebes at War: A Novel of Ancient Egypt* (New York: Anchor Books, 2005), originally published as Kifah Tiba, 1944. Although his historical details are sometimes questionable at best, as a storyteller, he recognizes that Seqenenre's life and death are not the end of story.

For Merneptah, Set-worshiping Apophis is not a partner of Egypt but a foe. If the full length of the story had been found it would have included a flood following the dialog between the Egyptian and the sleep-deprived Hyksos leader over noise as in the Mesopotamian tale of Atrahasis:

> Enlil heard their noise
> and addressed the great gods:
> The noise of humankind has become too intense for me,
> with their uproar I am deprived of sleep.[31]

The resulting noise-induced flood instead of proving Egypt was the gift of Nile, wreaked havoc in the land, turning Egyptian values topsy-turvy. Apophis was a false Horus who needed to be defeated so ma'at would be restored as it had been by Ahmose in The Tempest Stele.[32] In Merneptah's story, he claimed to have done so too. The humiliation of Seqenenre/Ramses by the Hyksos had been avenged. The seed of the Hyksos had been destroyed. Ramses had been defeated by his Apophis, Merneptah claimed to have defeated his.[33]

Like Nafouz, Merneptah drew on the traditional Theban image of the Hyksos but combined it with the events of his father's reign. The initial legacy of the Hyksos certainly was not a good one in Egypt. I will dispense with Manetho's depiction of the Hyksos. He wrote long after the time period in question and his description bears no resemblance to what the Hyksos actually did. His attempt to retroject the devastations and degradations of the Assyrian and Persian conquests sheds no light on the actual Hyksos 'invasion' or their legacy until Persian times. But the Assyrians, Persians, and Manetho were all centuries in the future for Merneptah. In his time, he looked to the legacy of the Theban interactions with the Hyksos and not the Delta-experiences of his father and grandfather.

In the Theban tradition. There also are multiple attestations of the Hyksos depicting a distinctly different image than the one on the Four Hundred Year Stela.

1. Seqenenre with his five-holed head undoubtedly did not share Ramses's positive perception of the Hyksos.[34]
2. Kamose with his stele also reflects a time of military confrontation with the Hyksos.
   (A) ruler is in Avaris, another in Kush,
   I sitting joined with an Asiatic and a Nubian,

31 Arnold and Beyer, *Readings from the Ancient Near East*, 26.

32 For Ahmose and the Tempest Stele see E. N. Davis, 'A Storm in Egypt during the Reign of Ahmose,' in *Thera and the Aegean World III: Chronology: Proceedings of the Third International Congress, Santorini, Greece, 3-9 September, 1989* (London: Thera Foundation, 1990), 232-235; Peter Feinman, 'The Tempest in the Tempest Stela: A Cosmic Story in History,' *Bulletin of the Egyptological Seminar* 19 (2015): 253-262; Karen Foster and Robert Ritner, 'Text, Storms and the Thera Eruption,' *Journal of Near Eastern Studies* 55 (1996): 1-14; Ryholt, The Political Situation in Egypt, 144; Robert K. Ritner and Nadine Moeller, 'The Ahmose 'Tempest Stela', Thera and Comparative Chronology,' *Journal of Near Eastern Studies* 73 (2014): 1 -19; Claude Vandersleyen, 'Une Tempête sous le regnè d'Amosis,' *Revue d'Égyptologie* 19 (1967): 123-159; Malcolm H. Weiner and James P. Allen, 'Separate Lives: The Ahmose Tempest Stela and the Theran Eruption,' *Journal of Near Eastern Studies* 57 (1998): 1-28.

33 The traditional Egyptological view of the story as a 'shaggy dog story' with only a vague recollection of the historical period in which it is set is comparable to criticizing Spartacus and MASH for inaccuracies about Rome and the Korean War. They were all written about their present although set in the past; see Donald B. Redford, *Egypt, Canaan, and Israel in Ancient Times* (Princeton: Princeton University Press, 1992), 108.

34 For the death of Seqenenre, see R. L. ten Berge and F. R. W. Van de Goot, 'Seqenenre Taa II, the Violent Death of a Pharaoh,' *Journal of Clinical Pathology* 55/3 (2002): 232; Feinman, 'The Quarrel Story,' 96-98; Garry J. Shaw, 'The Death of King Seqenenre Tao,' *Journal of the American Research Center in Egypt* 45 (2009): 159-176.

each man having his (own) portion of this Egypt,
sharing the land with me.[35]

The Theban-based dynasty sought to re-assert its dominion over the land. Although he did not succeed in banishing the Hyksos from the land, he did successfully change the battle lines by establishing a northern front.

3. Ahmose marked the full resumption of home-grown Theban-based rule over Egypt. (8) Then the gods [caused] the sky to come in a tempest of r[ain], with darkness in the western region and the sky being (9) unleashed without [cessation, louder than] the cries of the masses, more powerful than ....., [while the rain raged (?)] on the mountains louder than the noise of the (10) cataract which is at Elephantine. Every house, every quarter that they reached [...] (11) floating on the water like skiffs of papyrus opposite the royal residence for a period of [...] days, (12) while a torch could not be lit in the Two Lands.[36]

His Tempest Stele attests that all was right in the heavens and on earth. Ma'at had been restored.

The image of the Hyksos then abandoning Egypt lock-stock-and-barrel leaving the land Hyksos-free is a false one. Contrary to the image of an expulsion from Egypt, the Hyksos remained in the land for centuries to come. Manfred Bietak takes Egyptologists to task for ignoring the impact of the Hyksos, of tens of thousands of West Semites in Egypt for centuries.

> The end of Hyksos rule in Egypt from the historical point of view is a subject rarely addressed in Egyptology....In Egyptology, the impact of Hyksos rule on Egypt has been largely neglected in research if not ignored....[I]t is only logical to postulate that the presence of several ten thousands people of Western Asiatic people in north-eastern Egypt over a period of over 300 years (c. 1830-1530 BCE) must have had an impact on successive New Kingdom culture. It is highly unlikely that such a long time span of intense interaction between Egypt and a foreign population did not leave any traces.[37]

They had to have left traces. If those several tens of thousands had left the Egypt after 300 years, then truly it would have been an Exodus. There is no archaeological record of any such departure.

The Hyksos presence in Egypt did not cease after 1530 BCE. Just because the leaders of a military aristocracy have been displaced does not mean all the warriors are killed. Hyksos continued on in Egypt especially in the military. Bietak observes that the information from Manetho via Josephus contributed to the firm conviction among Egyptologists that the Hyksos had been driven out and returned to their own homeland across the river Nile.[38] The actual situation was

---

35 Quoted in Robert K. Ritner and Vincent A. Tobin (translators) in. *Literature of Ancient Egypt: An Anthology of Stories, Instructions, Stelae, Autobiographies, and Poetry* (ed. William Kelley Simpson; New Haven: Yale University Press, 2003), 346. For Kamose and the stela, see Labib Habachi, *The Second Stela of Kamose and His Struggle Against the Hyksos Ruler and His Capital* (Glückstadt: J. J. Augustin, 1972); T. G. H. James, 'Egypt: From the Expulsion of the Hyksos to Amenophis I,' in *The Cambridge Ancient History 2/1: The Middle East and the Aegean Region, c.1800-1380 BC* (ed. I. E. S. Edwards, et al.; Cambridge: Cambridge University Press, 1973), 289-312, here 289-299; H. S. Smith and A. Smith, 'A Reconsideration of the Kamose Texts,' *Zeitschrift für Ägyptische Sprach und Altertumskunde* 103 (1976): 48–76; Anthony Spalinger, 'Two Screen Plays: 'Kamose' and 'Apophis and Seqenenre," *Journal of Egyptian History* 3 (2010): 115–35.

36 Translation by Robert K. Rittner in Foster and Rittner, 'Texts, Storms, and the Thera Eruption,' 11.

37 Bietak, 'The Aftermath of the Hyksos in Avaris,' 20-21.

38 Bietak, 'The Aftermath of the Hyksos in Avaris,' 20.

not that stark. While some did leave, others remained. They became part of internal politics in Egypt, Delta-based warriors in a Theban-ruled country flexing its military muscle.

Hatshepsut's vigorous denunciation of the Hyksos in the Speos Artemidos Inscription needs to be put in its political context.

> The shores are united under my supervision, the Black Land and the Red Land under terror of me, (12) my impressiveness making foreign lands bow down. For the uraeus on my front pacifies for it all lands < ... >. (36) I do not sleep forgetting, (but) have made form what was ruined. For I have raised up what was dismembered beginning (37) from the time when the Asiatics were in the midst of the Delta, (in) Avaris, with vagrants in their midst, (38) toppling what had been made. They ruled without the Sun, and he did not act by god's decree down to my (own) uraeus-incarnation. (Now) I am set (39) on the Sun's thrones, having been foretold from ages of years as one born to take possession. I am come as Horus, the sole (40) uraeus spitting fire at my enemies. I have banished the gods' abomination, the earth removing their footprints.[39]

The reprimand of the warriors from across the river by the cobra Pharaoh seems excessive for a people who supposedly had been banished from history decades ago. One gets the impression that the Delta-based warriors were still present in the land and a potential threat to her Theban-based reign. Hatshepsut's words convey the opposite message Ramses would deliver over two centuries later.

Contrary to Hatshepsut's imprecation, the Hyksos appear to have assimilated into the new imperial order. In this connection, the figure of Aper-el is an intriguing one. He was vizier to Amenhotep III, vizier and instructor to Amenhotep IV, first servant to Aten, father of Huy, general of the chariotry and scribe of the recruits. Egyptologist Alain Zivie who excavated Aper-el's tomb, contends that Aper-el was Semitic based in particular on the 'El' component of his name. Zivie's claim probably is neither provable nor unprovable. Given Aper-el's position, if he was Semitic, he is likely to have been Hyksos. How exactly one would differentiate between an assimilated Hyksos in the land after two centuries and a pure Egyptian is problematical anyway. It also seems unlikely that a contemporary career military officer like Horemheb would not have been aware of and known Aper-el. Whether Aper-el's legacy continued into the 19th Dynasty of military officers cannot be determined.[40]

Another factor for consideration in analyzing the Egyptian military in the time of Ramses is the n'rn. The Egyptian army in the famous Battle of Kadesh included a division of n'rn. Not only did this division participate in the battle, they appear to have rescued the head-strong commander in-chief who marched into a Hittite trap and barely escaped with his life. Naturally, Ramses who was very meek, more than all men that were on the face of the earth, was eager to give credit

39 James P. Allen, 'The Speos Artemidos Inscription of Hatshepsut,' *Bulletin of the Egyptological Seminar* 16 (2002):1-17, pls.1+2; Alan H. Gardiner, 'Davies' Copy of the Great Speos Artemidos Inscription,' *Journal of Egyptian Archaeology* 32 (1946): 43-56; Hans Goedicke, *The Speos Artemidos Inscription of Hathshepsut and Related Discussions* (Oakville, CT: HALGO, Inc., 2004).

40 A. P. Zivié, 'Le Tresor Funeraire du Vizir 'Aper-el,' *Bulletin de la Société Française d'Égyptologie* 116 (1989): 31-44; Alain P. Zivié, *Découverte à Saqqarah - Le vizir oublié.* (Paris: Seuil, 1990); Alain P. Zivié, "Aper-el, Taouret et Houy: La fouille et l'enquete continuent,' *Bulletin de la Société Française d'Égyptologie* 126 (1993): 5-14; Eugène Strouhal, 'L'étude anthropologique et paléopathologique des restes du vizir 'Aper-el et sa famille: premiers résultats.' *Bulletin de la Société Française d'Égyptologie* 126 (1993): 24-33.

where credit was due, praised the n'rn for saving him. At least once he did when he was not taking credit for having won the battle all by himself.[41]

Generally, Egyptologists consider the n'rn to have been a division of West Semites. If so, then which West Semites? Whether the n'rn consisted of descendants of Canaanites who had migrated in to the land of their own free will over the centuries, captives from the imperial campaigns of Pharaoh into Canaan, or some combination of both is not the issue here. My question is: who were their commanding officers? There was no Canaanite army. Canaanites fought at the city level. There was no king of Canaan and the biblical reference to one at Hazor was to a city which suffered a brutal anti-Egyptian assault during the reign of Ramses.[42] Who among the West Semites had command experience and a military background to supervise a Canaanite division in the army of Ramses? The best answer is the Hyksos. I suggest the Hyksos commanded the n'rn division of Canaanites who rescued Ramses at Kadesh in year 5.

Hans Goedicke dates the Four Hundred Year Stele to shortly after year 34 in the reign of Ramses.[43] He asks:

> Why should Ramses in the second half of his reign suddenly have an urge to foster the legitimacy of his rule and that of his family, after they had occupied the throne for more than fifty years?

Donald Redford dates the dedication to 1257 BCE. He also claims that after the earlier battle of Kadesh:

> Headmen of Canaanite towns, vassals of Egypt, were impressed by what they divined as inherent weaknesses in Pharaoh's forces: poor intelligence and a tendency to panic. Rebellion was possible; Egypt *could* be beaten....In the wake of the retreating Egyptians, all Canaan flared into open revolt....It was Ramesses's darkest hour.[44]

41 For the n'rn, see Hans Goedicke, 'Considerations of the Battle of Kadesh,' *Journal of Egyptian Archaeology* 52 (1966): 71-80; Kitchen, *Pharaoh Triumphant*, 60; Colleen Manassa, *The Great Karnak Inscriptions of Merneptah: Grand Strategy in the 13th Century BC* (Yale Egyptological Studies 5; New Haven: Yale University Press, 2003), 53; Ellen Morris, *The Architecture of Imperialism: Military Bases and the Evolution of Foreign Policy in Egypt's New Kingdom* (Leiden: Brill, 2005), 362-366; Claude Obsomer, 'La Bataille de Qadech de Ramsès: Les n'arin, sekou tepy et questions d'itinéraires,' in *De la Nubie à Qadech: La Guerre dans l'Égypte/From Nubia to Kadesh: War in Ancient Egypt* (eds. Christina Karlshausen and Claude Obsomer; Brussels: Safran Publishers, 2016), 81-168; Alan Schulman, 'The N'RN at the Battle of Kadesh,' *Journal of the American Research Center in Egypt* 1 (1962): 47-52; Alan Schulman, 'The N'RN at Kadesh Once Again,' *Journal of the Society for the Study of Egyptian Antiquities* 11 (1981): 7-19; Anthomy Spalinger, 'Notes on the Reliefs of the Battle of Kadesh,' in Hans Goedicke, ed., *Perspectives on the Battle of Kadesh* (Baltimore: Halgo Inc., 1985), 1-42, here 3; Omar Zudhi, 'Benteshina and the N'rn Division,' *Journal of the Society for the Study of Egyptian Antiquities* 8 (1977-1978): 141-142.

42 For the destruction of Hazor during the reign of Ramses II see, Amnon Ben-Tor, 'The Fall of Canaanite Hazor-The 'Who' and 'When' Questions,' in Seymour Gitin, Amihai Mazar and Ephraim Stern, ed., *Mediterranean Peoples in Transition: Thirteenth to Early Tenth Centuries BCE* (Jerusalem: Israel Exploration Society, 1998), 456-467; Amnon Ben-Tor, 'Hazor and the Chronology of Northern Israel: A Reply to Israel Finkelstein,' *Bulletin of the American Schools of Oriental Research* 317 (2000): 9-15; Amnon Ben-Tor, 'The Sad Fate of Statues and the Mutilated Statues of Hazor,' in Seymour Gitin, J. Edward Wright, and J. P. Dessel, ed., *Confronting the Past: Archaeological and Historical Essays on Ancient Israel in Honor of William G. Dever* (Winona Lake: Eisenbrauns, 2006), 3-16; Amnon Ben-Tor and Sharon Zuckerman, 'Hazor at the End of the Late Bronze Age: Back to Basics,' *Bulletin of the American Schools of Oriental Research* 350 (2008):1-6; Piotor Bienkowski, 'The Role of Hazor in the Late Bronze Age,' *Palestine Exploration Quarterly* 117 (1987): 50-61; Israel Finkelstein, 'Hazor at the End of the Late Bronze Age,' *Ugarit Forschungen* 37 (2005): 341-349; Kenneth A. Kitchen, 'Hazor and Egypt: An Egyptological and Ancient Near-Eastern Perspective,' *Scandinavian Journal of the Old Testament* 16 (2002): 309-313.

43 Goedicke, 'Some Remarks,' 24.

44 Redford, *Egypt, Canaan, and Israel in Ancient Times*, 185.

Redford limits this awareness to Canaanites in the land of Canaan. Redford is correct about Canaanites revolting in the land of Canaan following Ramses's poor performance as commander in chief. Hazor is simply the most prominent example of the 'Canaanite spring,' the unrest Ramses now had to face in land of Canaan.

Meanwhile, all was not quiet on the home front. As Thomas Thompson astutely comments on the significance of the battle of Kadesh beyond the battle itself.

> After this defeat, Ramses II's army was racked with revolts. It had borne the brunt of the cost of his expensive misadventure....Civil unrest and religious opposition at home was doubly encouraged....A series of plots and intrigues by court factions bitter over the military failure at Kadesh effectively paralyzed royal authority and its control of import groups within the army.[45]

One might take issue to the extent to which unrest and intrigue occurred, but the basic thrust of the observation appears valid. Kadesh exposed the shortcomings the leader of the country and people responded to that weakness. Thompson has honed in on the precise time when the potential for disruption of ma'at in the political arena had occurred.

It is reasonable to conclude that Canaanites in the land of Egypt were equally capable of recognizing the shortcomings of the Egyptian commander-in-chief especially if they had rescued him in the battle of Kadesh. Canaanites in the Delta could revolt as well. Not all of them, just some. Perhaps led by a Hyksos Apophis who was very great in the land of Egypt, in the sight of Pharaoh's servants and in the sight of the people. Instead of a battle between the Hyksos in the north and the Thebans in the south, the battle had shifted to one within the capital city the Hyksos and the 19th Dynasty shared. These Hyksos and their allies lost of course since Ramses obviously was not removed from the throne. What did the Canaanites in the land of Egypt do after their revolt against Ramses failed? Did they disappear from history? Exactly what documentation would be required to show that Ramses had overcome a rebellion at home led by a Hyksos Apophis who was aware of Pharaoh's leadership shortcomings after the Battle of Kadesh?

The closest text I can find consistent with a Ramses-Apophis confrontation is Leiden Hymn 30.

The harpoon is deep in Apophis, the Evil,
        he falls by the sword;
    and those who chose war are huddled for slaughter--
Death cuts the hearts of God's demon enemies,
    who groan as outlaws,
        apostate forever....
He has ridden the waves unscathed
    and the rebels are no more!...
You have won, Amun-Re![46]

---

45 Thomas L. Thompson, *The Mythic Past: Biblical Archaeology and the Myth of Israel* (New York: Basic Books, 2000), 153.

46 John Foster, *Echoes of Egyptian Voices: An Anthology of Ancient Egyptian Poetry* (Norman, OK: University of Oklahoma Press, 1992), 67.

It is a water-based victory song over the same-named character who will be defeated in Merneptah's water-based Quarrel story a generation later. Perhaps this hymn from around year 52 sings to the demise of the Hyksos Apophis who had challenged his rule: he is dead and will never threaten Egypt again. But the anti-Egyptian people he had created did not die. They continued to be a threat thereby necessitating Merneptah's claim to have finished the job. The Hyksos who created Israel became the villains of Merneptah's Apophis story a generation after Ramses's Four-Hundred Year Stele when he reached out to Hyksos support.

Historical Reconstruction

The Four-Hundred Year Stele reflects a transitional point in time in the ebb and flow of the Hyksos within Egypt. They had been welcomed in the Delta and foes to Thebes. When a Theban Dynasty unified the country, some Hyksos left and some remained along with other Canaanites in the Delta. These warriors represented another power base to Pharaoh within the kingdom just as the priests of Amun did in the south. Hyksos assimilated into the Egyptian army primarily in the chariot force and reached the highest levels of power with Aper-el, just short of becoming Pharaoh himself. In the aftermath of the Amarna Age and for a short time only, military prowess not dynastic heritage became the prerequisite for becoming Pharaoh. It was even possible for a Hyksos to imagine taking the next step and become king in Egypt as an insider.

Ramses's failure at Kadesh was a signal to people throughout the land of Canaan and the Delta that Seti had not chosen wisely, he had not chosen the best qualified person for the job. A Hyksos 'Apophis' led and lost a rebellion against Ramses to replace him as king. The Hyksos 'Apophis' tried again this time in the name of Yahweh, the deity of the wilderness people ranked as lowest of the low in deity power. 'Yahweh took thee out of the land of Egypt' into the wilderness is the ultimate anti-ma'at topsy-turvy declaration. The warriors from across the river and others became the people Israel. They were an anti-Egyptian people who sought to liberate the land of Canaan from Egyptian hegemony with the support of El Canaanites and Yahweh Shasu.

Ramses fought to maintain control in the land of Canaan and only few cities joined the revolt. He campaigned in the land. He tightened his grip. He entered into an alliance with the Hittites. He married a Hittite princess who was not murdered on her journey to Egypt. He entered into an alliance with the Hyksos who had remained in Egypt and had been there 400 years. Israel remembered the 400 years Ramses had used and it became part of the cultural heritage of the Hyksos who did not remain in Egypt.

Chinese Sovereign Revolution: Temporal acceleration toward a better future?*

Maria Adele Carrai

A new rendering of 'revolution'

> 天地革而四時成，湯武革命，順乎天而應乎人，革之時義大矣哉！
>
> Just as Heaven and Earth make use of Radical Change so that the four seasons come to pass, so did Tang and Wu bring about Radical Change in the mandate to rule in compliance with the will of Heaven and in accordance with the wishes of mankind. A time of Radical change is great![1]

It is only at the very end of the Qing dynasty that China started to be conceived, both within China and abroad, as a sovereign or semi-sovereign state. Although its name, 'central kingdom,' *zhongguo* 中國, was maintained, by the end of the 19th century its perceived geographical and cultural centrality/superiority begun to lose credibility.[2] With it, the cosmological apparatus that legitimized the supreme authority of the Chinese Emperor started to crumble, reaching a culmination with the First-Sino Japanese War (1894-1895), which transformed Korea, one of the most loyal supporters of the Chinese imperial tribute system, into a sovereign entity obliged to obey the rules of international law rather than the cosmological ritual scheme imposed by the Chinese self-contained universe. The entire philosophical system that supported the traditional Chinese idea of imperial authority and history was deeply shaken; the ideological vacuum it created paved the way for the Chinese sovereign revolution of 1911. The political changes that took place in the year 1911, when the millennial continuity of the imperial system formally ended, absorbed China into modern history and implied a conceptual shift in the way that Chinese intellectuals spatially and temporally understood their country. This, it is argued, can be best understood by looking at the transformation of the meaning of the term *geming* 革命 in the Chinese context.

The word *geming* 革命 was not a neologism. It had appeared already in the *Classic of Changes, Yijing* 易經, a divination text of the Zhou dynasty (1046-256 BC) that exerted a great influence on Chinese thought, becoming one of the Confucian Classics in the 2nd century BC. *Geming*, change of mandate or great change, has often been understood as a violent but legitimate (in that it was divinely justified), change of mandate, and it was associated with the term *fanfa* 放

* This paper was first presented at the 2014 meeting of the Western Branch of the American Oriental Society at Stanford University.

1 Tang and Wu are great Chinese dynastic founders, the translation is from Richard John Lynn and Bi Wang, *The Classic of Changes: a new translation of the I Ching as interpreted by Wang Bi* (New York: Columbia University Press, 1994), 445.

2 It was quite emblematic that Japan, well aware of the symbolic value of the term 'middle kingdom,' at the end of the 19th century stopped using the term *zhonghua* 中華 to refer to China, and started to use the phonetic transliteration *zhina* 支那, which had a pejorative connotation. Ceng Yi (曾亦). 2012. 'Neiwai yu yixia- Gudai sixiang zhong de 'Zhongguo' guannian ji qi yanbian (内外与夷夏－古代思想中的'中国'观念及其演变),' in *Zhongguo zhi wei zhongguo* (中国之为中国)，Shanghai (上海): Shijijie chuban jituan (世界出版集团)，Fudan Daxue sixiangshi yanjiu zhongxin (复旦大学思想史研究中心), pp. 177, 199. 苗体君，窦春芳， 中国'支那'称谓的由来已变迁，in *Chinese Culture*, 2006,4, p. 7.

伐, which means the use of violence to overthrow a tyrannical monarch.[3] It is not a coincidence that in the passage from the *Yijing* quoted at the beginning of this section, the mythical figures of the emperors Tang and Wu are remembered. Both are famous in Chinese history because they carried through a *geming*, violently usurping the throne of the last kings of the previous dynasties.[4] The term was also used to refer to peaceful abdication; this was the case of the legendary King Yao, who abdicated in favor of the other legendary figure, King Shun. In either cases the term *geming*, can be translated as 'change of mandate' justified by Heaven, in that, in order to be such, it has to abide by the laws of nature, exemplified by the course of seasons, as well as human wishes, so long as they conform to Heaven's will. Thus a new ruler and a new dynasty can come into being through a change of mandate, and it is this regular and natural political change that defined Chinese cyclical history, understood, as I will later discuss, not so much as an evolution toward an unknown but better future, but rather as a repetition of the past, as long as it conforms with the rule of nature.

In the 19th century, under new pressures coming from Western powers, Chinese scholars started to be more receptive to Western thought, clearly shown in the increase in the number of translations of Western scholarship.[5] This process of translation and assimilation can be better understood by using the idea formulated by Lydia Liu of translingual practice; a process through which concepts and words are translated, adopted and appropriated in other languages. Liu emphasizes the fact that in translingual practice, the gradual legitimization of a new word and concept in a given host language takes place in an arena where there are constant struggles of a political and ideological nature to assert different interests. Thus the meaning of a translated term is a hybrid within the host language rather than a perfect equivalence between the two languages.[6] Not only were neologisms coined in order to accommodate new Western ideas, but often new concepts were introduced through the use of old terms, which were suddenly loaded with new meaning. This was the case in the Western notion of 'revolution', which was rendered in Chinese with the term *geming*. As Chan Jianhua demonstrated, *geming* not only started to be fused with the notion of rebellion, but it also started to mean a new departure for China in history; it was intimately related to the process by which China was becoming a new entity, acquiring a new sovereign subjectivity – the revolution or radical change was no longer governed by the constant principle of change dictated by *Tian* ('Heaven'), it became instead a self-legitimizing act, made in name of the new people, sovereign of a new nation.[7] In this sense, Chinese intellectuals needed to conceptualize such a transformation through a modern self-legitimizing revolution in order to get rid of China's past and its cyclical historical trajectory. This is evident

3 Chen Jianhua,《「革命」的現代性——中國革命話語考論》[*The Modernity of 'Geming': Textual Studies of Revolution Discourses in Modern China*] (Shanghai: Shanghai guji chubanshe, 2000), 243.

4 Both Tang and Wu, appealing to the mandate of heaven, *geming*, overthrew the ruling dynasty to establish their own. Tang, the first king of the Shang dynasty, established the new dynasty by overthrowing Jie, the last ruler of the Xia Dynasty, who was accused of disseminating terror. Wu did the same with the last emperor Xin of the Shang (called in a pejorative way *zhou* 紂) and established the Zhou dynasty. The issue of the change of mandate of Wu and Tang, fomented an almost millennial debate in Confucian and neo-Confucian circles about the legitimacy of a violent usurpation of power in the name of justice, reflecting the constant concern over the conflict between crude power and moral justice. Chen Jianhua,「革命」的現代性——中國革命話語考論》[*The Modernity of 'Geming': Textual Studies of Revolution Discourses in Modern China*] (Shanghai: Shanghai guji chubanshe, 2000), 240-242.

5 See Federico Masini, 'The Formation of Modern Chinese Lexicon and Its Evolution, Toward a National Language: The Period from 1840 to 1898,' *Journal of Chinese Linguistic Monograph Series* 6 (1993).

6 Lydia Liu, *Translingual Practice* (Stanford University Press, 1995), 26. See also Chen Jianhua, 'Chinese 'Revolution' in the Syntax of World Revolution,' in *Tokens of Exchange: The Problem of Translation in Global Circulations* (ed. Lydia H. Liu; Durham: Duke University Press, 1999), 355-374..

7 Chen Jianhua,「革命」的現代性——中國革命話語考論》[*The Modernity of 'Geming': Textual Studies of Revolution Discourses in Modern China*], (Shanghai: Shanghai guji chubanshe, 2000).

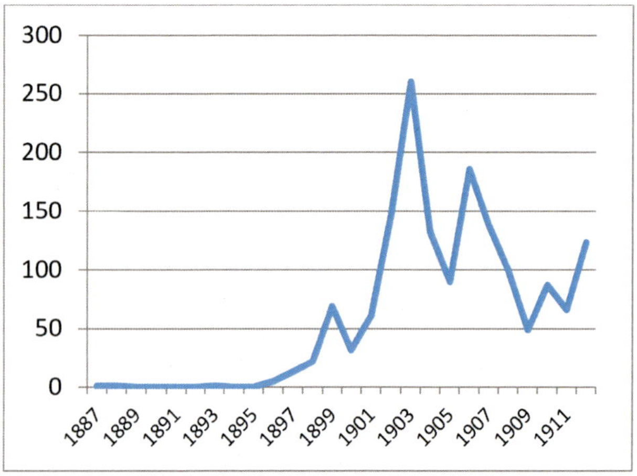

Chart 1. Use of the word geming 革命 in the years between 1887 -1912[9]

if we look at Chart 1 and the way the use of the word revolution, *geming*, increased dramatically after the First-Sino Japanese War, together with sovereignty *zhuquan*, independence *duli*, and equality *pingdeng*, reflecting the felt urgency not only for a radical change, but for a revolution in the modern sense, that would have emancipated China and transformed it from being an imperial divinely legitimized entity into a secular nation.[8]

Various intellectuals contributed to the transformation of the meaning of *geming*, and to the secularization and modernization of China. The aim of this paper is to look at the articulation of revolution through the work of some key thinkers at the end of the Qing dynasty and how they used the term revolution and other related concepts to promote the transformation of China into a sovereign nation. After having presented the main differences between the modern Western notion of revolution and the Chinese understanding of it in relation to the way history was constructed, the paper will proceed by looking at the confusion between revolution and rebellion introduced by Wang Tao, the first to discuss the French Revolution in his work. I will then discuss the importance of the introduction of the Western notion of evolution and progress by Yan Fu, and of the introduction of the notion of the people and the nation by Kang Youwei. I will further argue in the conclusion that these scholars contributed greatly to the paradigmatic shift embodied in the sovereign Revolution of 1911.

Geming: *From the space of experience to the horizon of expectation*

Before China defined itself as a sovereign nation, *Zhonghua minguo* 中华民国, in 1912, it was something different, in a way that is difficult for us to comprehend within the dominant Western nation-state paradigm. China was not a sovereign state in the modern sense and it did not fully participate in the Western international legal system, which at that stage was far from being inclusive of all nations.[10] China did not have a national flag, it did not have a national name and its imperial existence was divinely legitimized. In contrast to an agglomeration of border-defined states that were *superiorem non reconoscens* (did not recognize any superior) and that upheld the principle of equality among themselves, China upheld the principle of a hierarchically-organized unitary empire based on a differentiated cosmological metaphysics

8 Jin Guantao金观涛. *Guannianshi yanjiu:Zhongguo xiandai zhongyao zhengzhi shutu de xingcheng*, 观念史研究：中国现代重要政治术语的形成 [*A Research in Conceptual history: The formation of important political terms in China*] (Beijing: Falü Chubanshe, 2010), 386.

9 Data from the Full Text Database on the History of Modern Chinese Thought (1830-1930), established since 1997 by the Research Center for Contemporary Chinese Culture of the Chinese University of Hong Kong. This includes Chinese journals, newspapers, magazines, translations of missionary and Western works.

10 Gerrit Gong, *The Standard of 'Civilization' in International Society* (Oxford: Clarendon Press, 1984); Martti Koskenniemi, *The Gentle Civilizer of Nations* (Cambridge: Cambridge University Press, 2004).

through which its geopolitical, ritual and cultural centrality was justified.[11] Starting from the second half of the 19th century, this system underwent a deep crisis attributable to internal and external pressures. The tragic defeats during the Opium Wars left China with a loss of territory and sovereignty, but most importantly it challenged its claimed ontological centrality, creating a rupture in the cyclical vision of time that supported its millennial empire.[12] As a result, intellectuals and diplomats increasingly questioned the cosmological and philosophical apparatus that justified the Chinese imperial system, appropriating the new meaning of revolution and conceptualizing the political change they wished for their country in revolutionary terms.

If China experienced great changes, as defined in the *Yijing*, it never had a revolution similar to that of 1911, which apparently put an end not only to the Qing dynasty, but also to the millennial imperial dynastic system. This caused a dramatic disruption in the fundamental conception of time and nature, which were no longer uniform, but rather of a different quality, allowing Chinese intellectuals to project China's history towards a new, future horizon of perfectibility and salvation. 'Revolution,' as the German historian Reinhart Koselleck wrote, is one of the words that most naturally belong to the modern political vocabulary. While the term originally belonged to a physical-political order, in which 'the naturalistic metaphor of political 'revolution' lived on the assumption that historical time was itself of a uniform quality, contained within itself and repeatable,' and for which it was possible to conceive history and its always-valid paradigmatic examples, as *magistra vitae,*[13] with the French Revolution, the term started to be separated from its natural origin, becoming the '*fond mobile de la science humaine*', leading 'forward to the unknown future.'[14] Modern revolution, while interrupting, usually through violence, the political normality defined by a set of rules and values, modifies once and for all the relationship between what Koselleck identified as space of experience or the memory of the past (*Erfahrungsraum*), and the horizon of expectation or the expectation of the future (*Erwartungshorizont*). These are meta-historical categories that, in their constant intertwining, have always defined history and, in their changing relationship, have characterized different epochs. With modernity, for Koselleck, something new happened; the horizon of expectation expanded to the point of annihilating the space of experience, that is, the memory of the past. The future was from then on conceived always in positive terms as the horizon of perfectibility, in which history as *magistra vitae* cannot play any role. This sets free the '*historia humana,* which turned away from sacral history, and [leads to] the legitimation of a modern state capable of subduing salvation oriented religious factions.'[15] If revolution was legitimized by nature, in that it was related to astronomic phenomena, with modernity it became a self-legitimizing act.

Revolution, then, is inevitably linked to the modern idea of sovereign states, democracy and the atomistic views of sovereign individuals, who, freed from the bounds of the past and tradition, are unconstrained in formulating histories as progress, providing a first conceptual ground for the French Revolution.[16] As Claude Lefort remarked, the French Revolution is a democratic

11 For cosmology and the 'organising bipolarity' of the yin and yang and of the *wuxing* that founded the empire see for instance Angela Zito, *Of Body & Brush: Grand Sacrifice as Text/Performance in eighteenth-century China* (Chicago: University of Chicago Press, 1997), 149.

12 Ibid.

13 The quotation is from Reinhart Koselleck, *Futures past: on the semantics of historical time* (New York: Columbia University Press, 2004), 46. For space of experience as opposed to horizon of experience see ibid. 255-276, 310.

14 Koselleck, *Futures Past*, 49.

15 Koselleck, *Futures Past*, 16.

16 Richard Tuck,. *The Rights of War and Peace: Political Thought and the International Order from Grotius to Kant*. (Oxford: Oxford University Press, 1999); Francesco Gentile, *Filosofia del diritto. Le lezioni del quarantesimo anno raccolte dagli allievi*

moment in which the 'markers of certainty,' – that is, the legitimate basis for the production of political power and moral and legal rules, – are dissolved and with them also the normal relationship between the various corporate parts of society. This produces a great uncertainty in power and a projection into the future, which almost becomes a shelter.[17]

In China, the philosophy of history saw the historical fact as a natural phenomenon. In such a vision of the world, where social and political activities were but the extension of natural movements, the fundamental mechanisms of change did not, in principle, change. In Vandermeersch's words, in China 'a state apparatus...that voluntarily questioned the fundamental principles of the political organization of the society did not make sense. The idea that a variety of forms of government could exist had never crossed the mind of any Chinese author.'[18] This was, for instance, reflected in the idea of *wuwei* 無爲, non-action, of the Chinese sovereign, that is precisely that of the emperor who restores the natural order of things. It is not thus about his autonomy or self-legitimation, but about the spontaneity *ziran* 自然 of the integration between the human sphere and the normality/morality of the universe. 'The revolution, understood in the West as total liberation of man, it seems to the eyes of a Chinese puerile nonsense ignoring the true status of man in society and in the universe.'[19] The radical changes of the *geming* 革命, were meant to restore the just, moral, natural order; they were not revolutions in the modern sense, and the cosmological metaphysics that grounded the Chinese political order and the idea of centrality of the Chinese empire remained dominant until the end of the 19th century. It is not surprising, thus, that the first strong action taken by the Qing government in order to partly modernize its system after 1860, was called restoration, *tongzhi zhongxing* 同治中兴 ; it is only in the 1880s and 1890s that the tragic events of the First Sino-Japanese War, set up the ideological foundations for the sovereign revolution of 1911. From the collapse of the imperial order and the establishment of the *Zhonghua minguo* 中华民国, China immersed itself in the flux of modernity, projecting itself into an often utopian future, as in the case of the *Datong* imagined by Kang Youwei. Usually, China's clash with Western powers in the 19th century is believed to be the event that triggered the process of Chinese modernization. If this process, to some extent, should be considered one that characterized the entire history of China, the revolution of 1911 marked a paradigmatic shift in China, which generated a rupture from the historical, uniform quality that defined Chinese institutions under the cosmological ritual kinship framework for millennia.[20] In the next sections I look at some of the key intellectuals that contributed to the introduction and conceptualization of modern revolution in China and to the sovereign rupture.

Wang Tao the reformer who introduced the modern revolution

Wang Tao (王韜) (1828-1897), well versed in the Chinese tradition, was not a diplomat but a reformer, whose ideas had a strong influence on late Qing intellectual debate. He was the first to introduce the modern notion of revolution in China in 1890. He was a supporter of the Taiping rebellion, fighting for the fall of the Manchus, and for this reason he sought refuge in

(Padua: CEDAM, 2006).

17 Bernard Flynn, *The Philosophy of Claude Lefort* (Evanston: Northwestern University Press, 2005), 150.

18 Léon Vandermeersch,. ‹L'idée révolutionnaire, conception étrangère à la tradition chinoise: le 'changement de mandat' et la restoration de l'ordre cosmique,' *Extrême-Orient, Extrême-Occident* 2/2 (1983): 11-20.

19 Ibid., 19.

20 John Lagerwey, *China: a Religious State* (Hong Kong: Hong Kong University Press, 2010), 29. Etienne Balazs argued in relation to Chinese society during the Qing that it underwent 'a process of secularisation and bureaucratisation.' See Etienne Balazs, *Chinese Civilization and Bureaucracy* (New Haven, Yale University Press, 1964).

Hong Kong, where he helped James Legge translate the Chinese Classics. It is in Hong Kong that he started his career as a journalist. After having spent a period in Europe he returned to Hong Kong where, in the 1870s, he wrote two books that describe European power relations: *A Brief Introduction to France* 法國志略 and *Report on the Franco-Prussian War* 普法戰紀. The starting point of Wang reformist ideas is the unprecedented situation in which China found itself at the end of the 19th century, confronted by Western powers with superior military techniques. He asks rhetorically 'if China does not make any change at this time, how can it be on a par with the great nations of Europe and compare with them in power and strength?'[21] For Wang, the path of change is the path of reform and modernisation in the Western sense, and Wang's interest in Western history should be read in this sense, seen as a source of inspiration for Chinese reforms.[22]

If he removed China from what he percieved as its historical sinocentric isolationism and began to link the history of China with that of Europe and the world, he was not a revolutionary.[23] Although he introduced to China, for first time, the modern concept of 'revolution' in the preface of his *A Revision of the Outline of French History (Chongding Faguo zhilue* 重訂法國志略), published in 1890 and very popular among the reformers, Wang was foremost a reformist.[24] In this book, in which he discussed the history of France, he introduced the term 'revolution' by referring to the French Revolution. The work was mostly based on Japanese translations of texts on France and world history, and in particular on Okamoto Shiki's (1839-1904) *Historical Records of the World*, (*Wanguo shiji* 萬國史記), in which the idea of revolution appears in relation of the French Revolution. Here the word refers both to the greatest achievement of humanity, but also to a moment of great chaos and disorder, *luan*. This dichotomy between the two meanings, revolution and chaotic rebellion, was preserved by Wang, who, on the one hand condemned the French revolution for the chaos it created, but on the other, eulogized it because of its democratic results – which should be extended to China. This same ambiguous approach applies when Wang describes the death of Louis XVI who was executed during the French Revolution. On one hand, he was critical of the disorder engendered by the masses, called *luanmin* and *luandang*, who used violence against the king. However, he also believed that it is right to take up arms and rebel against tyranny, and that Louis XVI created the justification for his own death.[25]

Moroever, if he introduced the concept of modern revolution, he did not use it much himself to promote reforms inside of China, and this is the reason why he has been labelled as a 'reformer' rather than revolutionary. In his idea of reform, Wang started to envision China as a strong nation, especially given that the world he had in mind started to be dominated by Darwinist theory: 'the world of today may be summed up in two words: profit *li* 利 and power *qiang* 强.'[26] This is the reason why Wang was disillusioned about the effectiveness of the newly introduced international law and international treaties, which were simply disregarded by strong countries.[27] 'If one country is strong and the other is weak the first will not be willing to adhere

21 William Theodore De Bary, *Sources of Chinese Tradition* (New York: Columbia University Press, 1960), 253.
22 De Bary, *Chinese Tradition*, 254.
23 Paul A. Cohen, 'Wang T'ao's Perspective on a Changing World,' in *Approaches to Modern Chinese History* (ed. Rhoads Murphey and Mary Clabaugh Wright; Berkeley: University of California Press, 1967), 148-49.
24 重訂法國志略, 光绪庚寅（1890）仲春淞, 5 juan, 30b.
25 Ibid.
26 De Bary, *Chinese Tradition*, 254.
27 Ref. to a period of disillusion, Lam Hok-Chung 林學忠, *Cong wanguo gongfa dao gongfa waijiao* 从万国公法到公法外交 [*From the Law of Nations to International Diplomacy*] (Shanghai: Shanghai Guji Chubanshe, 2009).

to a treaty agreement forever, while if the first is weak and the other is strong the first will not be able to adhere to it over the long-run.'[28] In this sense, for Wang, international law was not the solution, and instead of looking forward, for a modern sovereign revolution, he looked back to the models offered by Chinese tradition. This is not to say that China should not change, on the contrary, Wang in one of his essays called for radical reforms: *bianfa* 變法. However, contrary to one of his successors, Kang Youwei, it was not a radical reform that implied a revolution or a new horizon of expectation, but a return to the past.[29] Even if he admired the technological capacities of the West, underlying the radical changes there was a belief in Chinese spiritual superiority. The acquisition of western technology, as well as its international legal apparatus, including sovereignty, was simply an instrument, it did not represent the way: 'they cannot be called the basis for governing the state and pacifying the world. The way of Confucius is the human way. As long as humankind exists, the way will remain unchanged.' And the way was the traditional three bonds and the five relationships.[30] Thus, what he called for was a radical change grounded on the Way (*dao*) set forth by Chinese tradition: 'After a few hundred years the Way will achieve a grand unity. As heaven has unified the south, north, east and west under one sky, it will harmonise the various teachings of the world and bring them back to the same source,' which was Confucian Great Learning.[31]

His thought is still embedded in a cyclical vision of history, which made it impossible to conceive a sovereign revolution in China. The revolution in Wang's eyes was a return to the past rather than a projection into an unknown but better future. For Wang, the cycles of history and rules are dictated by a moral Heaven, not by the autonomy of man or state, who, 'superseding national fortune (*guoyun*)' and acting morally, can eventually be rewarded.[32] 'China in the future will reassert its supremacy in the world,'[33] but as a re-establishment of the past, of the constant eternally-valid *dao.* In his essay 'China's enduring superiority' while China is the historical embodiment of *dao*, and its superior civilization is a constant in history, the West and its system of sovereign states is doomed to destruction.[34] Wang, nevertheless, contributed to the introduction of the modern meaning of revolution in China, unleashing its translingual power.

Yan Fu: The mandate of an evolutionary Heaven

In the Chinese tradition there was a deep correlation between the traditional idea of *geming* 革命, change of mandate, and *tianming*天命, Heaven's mandate. In order to be legitimate the *geming* has to find its justification in what can be considered as a sort of divine will of Heaven. The laws of change given by Heaven are repeatable and constant like nature, and as expressed in dynastical cyclical change they predict the cyclical nature of Chinese history. Thus, for China to enter as a nation into the modernized historical process it is also necessary to find a different way of defining Heaven. A key figure was the interpreter and translator trained in the Greenwich Naval College of London, Yan Fu 严复 (1854-1921).[35] Besides translating the works of John Stuart Mill, Herbert Spencer and Adam Smith, and Darwin's *Origin of Species*

28 Cohen, 'Wang T'ao's Perspective,' 142.
29 In Taoyuan wenlunwaibian 1:11-15b, translated in De Bary, *Chinese Tradition*, pp. 252-254.
30 Ibid., 253.
31 Ibid.
32 Cohen, 'Wang T'ao's Perspective,' 150-151.
33 Ibid., 153.
34 Ibid., 153-154.
35 Benjamin I. Schwartz, *In Search of Wealth and Power: Yen Fu and the West.* (Cambridge, Mass.: Belknap Press, 1964).

(translated as *Wuzhong yuanshi* 物種原始) in 1906, Yan Fu was the first to translate Huxley's *Evolution and Ethics,* in 1898. The idea of nature herein expressed was crucial because it allowed Chinese intellectuals to think differently about time and nature. Nature, for Huxley, does not have a moral value,[36] and thus the idea that nature is moral, which served to underpin the philosophy of history as *magistra vitae*, according to which examples from history can be taken as a constant model for life and law, cannot be supported. The logic of the repetition of the past is replaced by that of natural selection and historical progress. This idea of evolution also affected the notion of Heaven, *tian*. While up until this point the moral *tian*, *tianyi de tian*正義的天, legitimized the cosmic Chinese imperial order, from the beginning of the 20th century, *tian* became the principle of evolution, *tianyan de tian* 天演的天, that rewarded the strongest or the fittest.[37] With such a change in the understanding of *tian*, the ultimate guarantor of the cosmic imperial order was dismissed. There is no space for a virtuous Son of Heaven in a world where the strongest is rewarded, and this was clear to many intelletcuals of the late Qing period, which saw an increasingly powerless government incapable of facing Western incursions. The new evolutionary *tian* fully legitimizes the sovereign revolution in a globe dominated by 'the reality of a cruel law without justice or humanity.'[38] From the late 1890s, international society and international law were no longer an option, but a necessity and an overwhelming reality. They were increasingly understood as systems in which it was no longer the morality of the *common law* 公法, to determine the conduct of states, but rather their material strength. The way Chinese intellectuals conceptualized 'sovereignty', and the naturalistic and universalistic discourse of international law that supported the first introduction of sovereignty in China, was substituted by a positivist one: 'In this period [early twentieth century] most people have already abandoned ideas of natural law, they do no longer only emphasize the idea of international law as natural reason or the law that every country respects, or the human ideal norm. They started to be influenced by legal positivism, and critics clearly started to point out that international law is the rule that governs the relations between civilized countries... what is emphasized is its contractual and precedent foundation, not reason's.'[39] The internal discourse of international law and sovereignty started to be dominated by the evolutionary idea of 'the strong eat the weak' (*ruorou qiangshi* 弱肉強食), China was seen as a weak country, *ruo* 弱, which in order to become *qiang* 强, had to dismiss its lame imperial system and initiate a sovereign revolution, which saw the masses as masters of history. It is not surprising, thus, to see the dramatic increase in the use of the term revolution in and after 1903; at the same time the terms directly associated with it of independence, *duli* 獨立 and equality, *pingdeng*,平等 also dramatically increased,[40] becoming the new, meagre source of morality, in a world where only the fittest survives.

Scholars like 'Yen Fu, Liang Ch'i-ch'ao, and even K'ang Yu-wei accepted the Western multi-state system with an astonishing lack of reservation. They were all willing to see the emperor reduced from his unique, cosmic status to the position of national monarch.'[41] These intellectuals,

36 So, Billy 苏基朗. 'You fa wu tian? Yan Fu yi 'tian yan lun' dui 20 shiji chu zhongguo falü de yingxiang 有法无天？严复译'天演论'对 20 世纪初中国法律的影响' [An immoral law? Yan Fu translation of the Evolutionary theory and its influence on Chinese law of the 20th century]. *Qinghua faxue* 清华法学 6，No. 5 (2012): 131

37 Ibid., 136.

38 Ibid., 137.

39 Lam Hok-Chung 林學忠, *Cong wanguo gongfa dao gongfa waijiao* 从万国公法到公法外交 [*From the Law of Nations to International Diplomacy*] (Shanghai: Shanghai Guji Chubanshe, 2009) 230.

40 Jin Guantao金观涛. *Guannianshi yanjiu:Zhongguo xiandai zhongyao zhengzhi shutu de xingcheng,* 观念史研究：中国现代重要政治术语的形成 [*A Research in Conceptual history: The formation of important political terms in China*] (Beijing: Falü Chubanshe, 2010), 386.

41 Schwartz, *Wealth and Power,* 285.

together with the revolutionaries and the reformists, started to link external and internal sovereignty, arguing that as long as the government was weak and the people did not have a consciousness of the nation, China could not rescue its sovereignty. In other words, China's revolution started to become a link to the people, which I discuss in the next section. The new motto was 'China is the China of the Chinese people'; this it was expressed through a racial discourse of national sovereignty (民族国家).[42]

Kang Youwei and the national revolution

Kang Youwei (康有为) (1858-1927), a controversial political thinker and supporter of the creation of a constitutional monarchy in China, contributed profoundly to the Chinese sovereign revolution and to the development of the modern Chinese national consciousness. The aim of his revolutionary political thought was to give China a new national collective identity, and in 1898 his ideas managed to influence the court decisions in the short-lived One Hundred Days Reform. The program that inspired the reform, the 30 articles of the Baoguohui zhangcheng 保国会章程 of April 1898, laid out a series of principles that, for the scholar Wang Ermin, constituted one of the earliest full expressions of modern nationalism.[43] In the program, Kang referred to three principles: protecting Chinese national sovereignty (art.3), protecting the independence of people (art. 4), protecting Confucianism (art. 5).

By looking at the database of the Chinese University of Hong Kong, it is possible to observe that Kang Youwei (1858-1927) used the word *geming* a few times. In 1887 he used it for the first time in his *Bibliography of Japanese Books*, to refer to the French Revolution, *Foguo geming* 佛国革命. The term re-appeared many times in his writings and Kang tended to condemn the violence of the French Revolution, but 'praised it for establishing a democratic nation state, the highest political achievement on earth.'[44] If he was aware of the notion of the modern revolution, however, he was cautious not to abuse it in his memorials to the Court, because of its implicit meaning, which would have made the current Manchu rule seem illegitimate. As Chen Jianhua demonstrated, Kang, instead of using the term revolution directly, imposed it indirectly by implying it in his memorials to the throne with the terms *bianfa* 变法 (changing the law), *weixin* 维新 (preserving the newness) and *bianzheng* 变政(changing the political institutions).[45] The first move in this direction was to transform Confucius into a reformer in his work of 1896, 'A Study of Confucius as a Reformer' *Kongzi gaizhi kao* 孔子改制考, in which Confucius was portrayed as the sage who changed the laws *bianfa zhi sheng* 变法之圣. Kang 'imposed on the Qing court the universal law of world revolution' meaning the horizon of expectation that made it possible to detach China from its past, imagining it as a constitutional monarchy.[46]

For Kang, China needed serious and profound political reforms in order to protect its sovereignty. One of the most important changes Kang supported was the introduction of democracy. The theoretical foundation of his thought can be found in his earlier works, written in the 1880s, *A Complete Book of Substantial Truths and Universal Principles*（實理公法全書）and *Esoteric and*

42 '中国者，中国人之中国' 'It is only through independent and autonomous people, who unite, that we constitute a true nation' '凡同种人务独立自治,联合统一以组织一完全之国家是也.' 国魂篇, 载 浙江潮, 第1 期(1903年 2 月)
43 保国会章程 https://zh.wikisource.org/zh-hant/保國會章程, 227.
44 Kang Youwei, *Ouzhou shi yi guo you ji.* (Shanghai: Guang zhi shu ju, 1906). http://catalog.hathitrust.org/api/volumes/oclc/35114332.html. 1906 On the History of French Revolution 法國革命史論 in xinmin congbao新民叢報. Ming Yi (Kang Youwei), Faguo geming shi lun, xinmin congbao 85 (august 1906) 87 (September 1906).
45 Chen Jianhua, 'World Revolution Knocking at the Heavenly Gate: Kang Youwei and His Use of Geming in 1898,' *Journal of Modern Chinese History* (London: Routledge), 5/1 (June 2011), p. 90.
46 Ibid., 91, 康有爲, 孔子改制考, 第1版 (Shanghai : Shanghai shu dian, 1992).

Exoteric Essays of Master Kang (康子内外篇). In his *Complete Book of Substantial Truths and Universal Principles* Kang argues that humans are equal and need to be so, in order to realise the universal and undifferentiated benevolence, *ren* 仁.[47] This equality is justified by the fact that 'Heaven and Earth's original substance was granted to man' '天地原質以爲人,' which also grants to each individual the 'right to autonomy' 自主之權 and equality.[48]

In his giving of autonomy and equality to people, locating sovereignty in the people, he anticipates the shift in the location of authority of his pupil Liang Qichao. Kang specifically saw parliamentary democracy, in which the sovereign is simply a mediator *zhongbao* 中保, as the most beneficial form of government for humanity. Kang was struggling to change China radically. In his conversation with the emperor in June 1898, he remarked that 'the requisites of [a needed] reform are that all laws and the political and social system be changed and decided anew.'[49] Although he attempted to protect Confucianism, at the same time he contributed to erode the sacrality of Chinese state doctrine, reducing it to one among many ideologies. The revolution Kang had in mind, in other words, was not aimed at preserving Chinese culture; its foremost objective was the preservation of China as a sovereign country, and the reformer Confucius had the role of facilitating this progress. The past he referred to was itself revolutionised, and manipulated in order to accommodate the present China's need to become a sovereign nation. If in his earlier works the idea of progress, although somehow present, was not yet completely articulated, in his later work, *Datongshu*, probably also because of the increasing influence of Darwinist and evolutionary theories introduced by Yan Fu, the notion of revolution dominated his philosophical apparatus.[50] In this book he immerses China in linear progressive history, describing three ages: the Age of Disorder (*juluan shi* 据乱世), the Age of Approaching Peace (*shengping shi*升平世), and the Age of Universal Peace (*taiping shi* 太平世).[51] China can no longer be an empire; it has to go through the phase of sovereign nation state to finally reach the Great Unity within the new linear vision of history.

Conclusion

> 我四万万同胞如新婴儿新出于母胎，从今日起为新国民，道德一新、学术一新、冠裳一新，前途种种新事业，胥吾新国民之新责任也.
>
> We four hundred million compatriots are like a new baby born from a new mother's womb, from today on we will form a new nation, a new morality, a new education, a new guise, promising all sorts of new business: this is our new responsibility for the new nation. [52]

The Xinhai revolution of 1911 was justified in terms of being new and for the way in which it broke away from the past, with the promise of a better future. In 1912, the editorial of the influential journal *Shenbao* reported above provides a snapshot of this new spirit, which was the result of the materialization of revolutionary ideas. Sun Yatsen, the first president of the Chinese

47 Kang, 686
48 Kang, 287.
49 Doc. 46, Ssu-yü Teng and John K. Fairbank, *China's Response to the West* (Cambridge Mass., Harvard University Press, 1954), 177.
50 Jung-pang Lo, 'K'ang Yu-wei and His Philosophy of Political Change and Historical Progress,' *Symposium on Chinese Studies: Commemorating the Golden Jubilee of the University of Hong Kong, 1911-1961* (University of Hong Kong 1968), vol. 3, p. 80. Xianggang da xue wu shi zhou nian ji nian lun wen ji : 1911-1961.
香港 : 香港大學出版社, 1964-1968. San pao li: 686.
51 Kang, pp. 327-337.
52 Xin Zhuci《新祝词》Shenbao《申报》(January 1, 1912)

Republic, together with his revolutionary party *genmingdang* 革命党, overthrew Manchu rule in the name of a revolution that would have established a new constitutional China.[53]

If China did not seem radically changed after the year 1911, – there was continuity with some of its economic and social institutions –, at an ideological level the year 1911 marked a crucial point in its history.[54] The Xinhai Revolution saw a change in the basic assumptions and paradigms that defined China until that point. Already, in the year 1905 with the abolition of the examination system, Chinese cultural continuity had been seriously threatened, but in the year 1911, as Chen argued, 'the self-justified revolution acted as a total sovereignty, and thus the Mandate of Heaven was transformed into an active authority, effectively mobilizing the masses by awakening their collective memory of the reborn kingdoms of ancient China.'[55] The imminence of Heaven produced a revolution in Chinese history, which was no longer legitimized by the constancy of the rule of nature as reflected in traditional cyclical dynastic rule, but that found its new justification in the Chinese people, new citizens of a sovereign China that alone was building its own future. This new dimension of temporality was clearly reflected in the change of the calendar, which ceased to belong to a particular dynasty. It was replaced by the Western calendar of linear temporality. This signaled a clear rupture with past; the affirmation of the dynastic cyclical past was no longer, instead it affirmed the present and future of China as a nation state inserted in the modern teleology.[56]

Wang Tao, Yan Fu, Kang Youwei, together with other intellectuals of the late Qing dynasty, struggled to redefine Chinese subjectivity and immerse it in the course of modern linear history. The Xinhai revolution brought about the abandonment of the self-contained Chinese universe. This, together with the clash with the world of the Western nation state, produced both utopic visions of the future and idealism which frequently faced disillusionment, and an often blind return to the past.

53 Sun Yat-sen (1866-1925), trained almost entirely abroad (he attended his secondary school in Hawaii, and medical school in Hong Kong) was a prominent spokesman of revolutionary thought and the new sovereign nation, becoming the first president of the First Republic of China in 1912. After having founded the the Nationalist Party, the Kuomintang, realising the uselessness of the Manchu reforms, in 1895 he attempted and failed a *coup d'etat* in Canton. He then joined a secret revolutionary society, supported by the Chinese in exile and students sent to Japan and formed the China *Tongmenghui* 同盟會 (League of United Alliance). Through their journal started in 1905, the 民報, they propagated their revolutionary ideas, and issued the manifesto of the Revolution where the Three People's Principles appeared. It is Sun that distinguishes between Manchu race and Manchu government, the Manchu race being part of China, answering, in his way, the racist nationalism that was emerging in China, directed against the Manchu race. Sung's Three Great Principles were originally directed against Manchu in 1905, but after 1911 they became the Three People's Principles.

54 Xu Guoqi. 'The Age of Innocence: The First World War and China's Quest for National Identity' (PhD Dissertation; Cambridge, Mass.: Harvard University, 1999), and Peter Zarrow, *After Empire: The Conceptual Transformation of the Chinese State, 1885-1924* (Stanford, California: Stanford University Press, 2012), 30.

55 Chen Jianhua, 'World Revolution Knocking,' 92.

56 Xu Guoqi, *Age of Innocence* 49-50.

Nippur's Galileo Problem: Time for a Revolutionary new Beginning to Time?

JoAnn Scurlock

In an article in JCS 68, Jonathan S. Tenney published a Middle Babylonian offering list from Nippur that, he argues, shows that the rise of Marduk to the head of the pantheon as reflected in the Babylonian New Year's Festival and the accompanying *Enuma eliš* was earlier than previously suspected and reflects the reception of Babylonian intellectual creations in Assyria via Nippur.[1] I am not averse to pushing back the composition of the Babylonian creation epic into the late Kassite period, nor indeed to the idea that Assyrian ritual was heavily influenced from Nippur as is clearly indicated in one of the surviving Assyrian cultic commentaries.[2] What I do have difficulty believing is that Nippur was the vector by which the celebration in the Babylonian New Years' Festival of Marduk's accession to the exalted position of King of the Gods reached Assyria. This is for one simple reason. Unfortunately for Tenney's argument, a late cultic commentary reveals, as we shall soon see, that Nippur never did, in fact, accept the rise of Marduk to the head of the pantheon.

At issue is the interpretation of cultic events that took place in the month of Nisannu. In the Nippur offering list published by Tenney, the offerings of the eighth of Nisannu were to be dispensed at Nippur for the *akītu* of the Temple of Marduk.[3] In the later Babylonian New Years' Festival, this day is associated with the return of Marduk to his temple from the *akītu*-house, presumably the event being honored at Nippur.[4] The high point of this festival is marked in the Seleucid ritual as the 5th of Nisannu. On this day, the *Enūma eliš* was read while the emblems of Anu and Enlil were veiled.[5] This ritual certainly signaled at Babylon Marduk's defeat of both Anu and Enlil and his rise to the head of the pantheon, particularly in combination with a bonfire sacrifice that was celebrated on that day and that was understood as representing Anu's incineration as follows:

'Forty minutes after sunset, the high priest ties forty straight reeds of three cubits each, which have been neither cut nor broken into a bundle with a palm frond. In the great courtyard, they open up a pit, and he puts (them) into the pit. He puts in honey, ghee, pure oil (and) ... They ... a white ox bef[ore the planet Mercury].[6] The king [introduces] an ignited fire into it by means

1 Jonathan S. Tenney, 'The Elevation of Marduk Revisited: Festivals and Sacrifices at Nippur during the High Kassite Period,' *Journal of Cuneiform Studies* 68 (2016): 153, 162-177.

2 '[The rites that] are performed [in the E]gašankalamma are just like those of Nippur' Alasdair Livingstone, *Court Poetry and Literary Miscellanea* (State Archives of Assyria 3; Helsinki: Helsinki University Press, 1989), no. 38: 2-3.

3 Tenney, 'Elevation of Marduk,' 164 obv. 12'.

4 See below.

5 François Thureau-Dangin, *Rituels accadiens* (Paris: Ernest Leroux, 1921), 136 lines 280-284 = Marc J.H. Linssen, *The Cults of Uruk and Babylon: The Temple Ritual Texts as Evidence for Hellenistic Cult Practises* (Cuneiform Monographs 25; Leiden: Brill-Styx, 2004), 219-220, 228 lines 280-284.

6 The name of the planet Mercury is *šiḫṭu* from *šaḫāṭu* for which the Sumerogram is GU_4.UD. The same signs taken another way, however, and the Sumerogram could mean 'white ox' (GU_4.BABBAR), hence the choice of offering.

of a reed. The king and [high priest say] this *naqbītu*-prayer: "Shining Mercury[7] that brigh[tens the darkness ...] burner of Anu.'"[8]

The view from Nippur as to the significance of this enacted cosmic event was quite different from the interpretation placed on it by the priests of Marduk. OECT 11 no. 69+70, a late commentary, reveals that the defeat of the 5th of Nisannu was indeed the defeat of Anu but by Enlil and the rise not of Marduk but of Enlil to the head of the pantheon as follows:

> '[In the month Ni]sannu, day five, when the *akītu*-festival of Marduk is held, [and a brush fire] is kindled [in] the gate of the Ekur, (it is) because Enlil entered the uppermost heaven and drew near [to ...], put on the crown of Anu (and) sat down on the throne of kingship, and Nusku was consecrated with him. DN sacrificed a [...] of greeting to Enlil (and) Enlil was raised to the rites of Anuship' (OECT 11 no. 69+70 i 3'-8')

To unravel this mysterious disjunction, it is necessary to have a closer look at our sources. There is a back story to the relationship between Anu, Enlil and Marduk encapsulated in the ritual for the Babylonian New Years' Festival and in the Babylonian Creation Epic itself which may be partially reconstructed with the help of late cultic commentaries as well as information gleaned from reconstructions of the festival calendar at Sumerian Nippur.

It is remarkable that, in the Seleucid Babylonian New Year's Festival ritual, the burning of Anu is carried out through the agency of the planet Mercury. If it had been done by Marduk for Marduk's own benefit, this defeat should really have been carried out through the agency of Marduk's planet Jupiter rather than Mercury, originally Ninurta's planet but taken over in Babylon by Marduk's son Nabû. What we are seeing here, I would argue, is a preserved fossil of an older ritual in which it was Ninurta, working for his father Enlil who disposed of his father's rival. If so, then the interpretation placed on it by the OECT 11 no. 69+70 commentary will have been the original understanding of that ritual in Nippur theology, stubbornly still cited long after Marduk's rise to King of the Gods status in Babylon.

The Babylonian New Years' Festival celebrates the annual defeat of the Forces of Order as represented by primordial stars[9] by the Forces of Chaos as represented by a planetary[10] warrior god armed with divine winds and working on behalf of his divine parents. Yes, you heard me, the Forces of Order were defeated by the Forces of Chaos in the *Enūma eliš* and it is to celebrate this victory that ancient Mesopotamians celebrated the Babylonian New Years' Festival. It is to be remembered that they defined Cosmic Order in terms of the Tablet of Destinies which were, at the time of Marduk's great victory firmly in the hands or rather more precisely around the neck of his opponent.[11] From the time that we first encounter him, baby Marduk embodies chaos in the form of the dust storms that he stirs up and the waves that he makes in his orderly greatgrandmother Tiamat's still cosmic waters, thus disturbing the primordial gods and

7 The interpretation follows Jeremy A. Black, 'The New Year Ceremonies in Ancient Babylon: 'Taking Bel by the Hand' and A Cultic Picnic,' *Religion* 11 (1981): 45, 51 but it should be noted that the copy has a clear ZÁLAG in line 461 where an UD would be required.

8 Thureau-Dangin, *Rituels accadiens* 145-146 lines 453-462 = Linssen, *Cults of Uruk and Babylon*, 223, 232 lines 453-462.

9 Enmešarra, about whom more presently, is a constellation called the 'Old Man'. For references, see *The Assyrian Dictionary of the Oriental Institute of the University of Chicago* Š/II (Chicago: The Oriental Institute, 1992), 398-399.

10 Marduk is associated with the planet Jupiter. For references, see *The Assyrian Dictionary of the Oriental Institute of the University of Chicago* N/2 (Chicago: The Oriental Institute, 1980), 147a.

11 *Enūma Eliš* I 157-158, edited by Philippe Talon, *Enūma Eliš: The Standard Babylonian Creation Myth*, State Archives of Assyria Cuneiform Texts; Helsinki: The Neo-Assyrian Text Corpus Project, 2005).

engendering the coming conflict.[12] Moreover, time is movement and movement chaos so that human beings as creatures of time are themselves forces of chaos (with small caps). In other words, the victory of the Forces of Chaos (with large caps) was also a victory for humanity, and good reason to be celebrating.

These ideas were hardly new with the Babylonian New Year's Festival, which nobody imagines to be earlier than the Old Babylonian Period. Before Marduk took on his starry ancestors there must have been a Sumerian hero who performed the same task and, indeed there was. From earliest times, Nisannu was the first month of the year at Nippur, the month when Enlil was enthroned, taking his place at the head of the divine assembly.[13] Nisannu [14]was not, however, the first month of the first year ever. This was celebrated, as later, at the Vernal Equinox. And when the Nippur cultic calendar reached final form, the Vernal Equinox was in Ayyaru rather than Nisannu.[15] The reasons for this disjunction between the beginning of time and the beginning of the calendar are not far to seek. Enlil was interested in power not creation, certainly not mankind's creation, as the Atra-Ḫasīs makes abundantly clear.[16] He was also no warrior, that task generally falling to his illegitimate son Ninurta,[17] the city god of Nippur whose *akītu* was in Ayyaru.[18]

Enlil's rise to prominence required first the defeat of rival Anu in Nisannu followed in Ayyaru by the doing-in of the elder generation of gods as understood at Nippur, a constellation called Old Man Enmešarra and his sons, the Pleiades. When lunar and solar years were lined up properly, the Pleiades' heliacal rising in Ayyaru would occur at what was, between approximately 4,000 and 2,000 BCE, the month of the Vernal Equinox.[19] The first ever appearance of the Pleiades and their first ever movement across the sky was, in Nippur's conceptualization, the result of the battle between them and Ninurta as represented by the planet Mercury which was imagined as having had its first ever solstice point in the first month of the first year ever. It was this battle that, so to speak, started the clock or, in other words, this was the genesis of linear time. Since humans are creatures of linear time and could not, in Mesopotamian's way of thinking, be alive in eternity, it was essential that the clock be kept running and it was for this reason that the annual festivals in the month assigned to the battles that created time were celebrated and not to somehow stop the clock as is all too often asserted. This latter notion is, if truth be told, not based on any evidence from ancient Mesopotamia but on Mircea Eliade's myth of the eternal return,[20] a myth indeed, and of his own making.

---

12 *Enūma Eliš* I 105-110.

13 Mark E. Cohen, *Festivals and Calendars of the Ancient Near East* (Baltimore: CDL Press, 2015), 117-119.

14 Hermann Hunger and David Pingree, *MUL.APIN: An Astronomical Compendium in Cuneiform* (Archiv für Orientforschung Beiheft 24; Horn: Ferdinand Berger & Söhne, 1989), 40.

15 The Vernal Equinox was in Taurus (= Ayyaru) between approximately 4,000 and 2,000 BCE. See Catalin Anghelina, 'On the Mythology of Oceanus,' *Journal of Ancient Near Eastern Religions* 9 (2009): 145.

16 *Atra-Hasīs* tells the story of the creation of mankind. Essentially the Anunnaki gods, who had been doing all the work, went on strike. Human beings were created to do all the work so that the Anunnaki could put their feet up, but they had to practically burn down Enlil's house to get him to agree to this, and even so, their ringleader got executed. Edited by Wilfred G. Lambert and Allan Millard, *Atra-Ḫasīs: The Babylonian Story of the Flood* (Oxford: The Clarendon Press, 1969), 42-65: I 1-305.

17 Maria Grazia Masetti-Rouault, 'Fathers and Sons in Syro-Mesopotamian Pantheons: Problems of Identity and Succession in Cuneiform Traditions,' in *La famille dans le Proche-Orient ancien : réalités, symbolismes et image: Proceedings of the 55th Rencontre Assyrioloque International, Paris 6-9 July 2009* (ed. Lionel Marti; Winona Lake (Eisenbrauns, 2014), 261 n. 30.

18 Oliver R. Gurney, *Literary and Miscellaneous Texts in the Ashmolean Museum* (Oxford Editions of Cuneiform Texts 11 [OECT 11]; Oxford: Clarendon Press, 1989), no. 69-70 i 32'-34'.

19 Angelhina, 'Mythology of Oceanus,' 145 n. 16.

20 Mircea Eliade, *The Myth of the Eternal Return: Cosmos and History* (Princeton: Princeton University Press, 1971). See the

Old Man Enmešarra was already long since visible in the sky in Ayyaru but, when encountering Ninurta, he was imagined as lying down in an Elamite chariot drawn by Anzû/Pegasus.[21] The reason that he was lying down was that Ninurta had killed him. The poor Old Man was done in by a classic piece of treachery; Enlil married his daughter, Anunitu,[22] and then sent his son Ninurta to kill Enlil's new father and brothers-in-law.[23] Parallels to the legend of Pelops and the prominent role of Anzû at the court of Enlil before his theft of the Tablet of Destinies[24] suggest that Anzû was persuaded by promise of reward to cause the chariot to which Enmešarra was hitched to crash.

Arriving on the scene as the upstart divinity of the Amorite rulers of Babylon, the god Marduk soon established a foothold in the Sumerian pantheon as son-in-law of the Sumerian chief god Enlil. To this end, a Sacred Marriage was performed between Marduk and Enlil's daughter, Ištar, a marriage still celebrated in the ritual of the Seleucid New Years' Festival. By the end of the Festival, Ištar had been divorced on grounds of adultery and Marduk was safely married to Ṣarpanitum. The details of this sordid affair shall hopefully be outlined in a planned future monograph.[25]

In any case, the father-in-lawship of Marduk survived the shenanigans of a sexually incontinent goddess. Marduk was to assist his new wife's half-brother Ninurta in the annual defeat of the sky god Anu which ensured Enlil command of heaven and earth and the right to place his throne above the stars. However bravely Marduk may have fought, however, the victory of someone who acts on the command of another is not his victory but that of the one who commissioned him. This meant that the fruits of the victory went to Enlil, who rose to the head of the pantheon on Marduk's shoulders. Enlil was not, of course, ungrateful and this is why Nippur celebrated, on the eighth of Nisannu, Marduk's return from the *akītu*-house where he had gone to celebrate his marriage to Ṣarpanitum.[26]

Nevertheless, when time as we know it was created by the productive slaughter of Enmešarra and sons, it was Ninurta and not Marduk who was Enlil's general. This is for the simple reason that Nisannu, the month of Marduk's empty triumph may have been the first in Nippur's calendar, but the all important battle with the stars that forced them to wake up from their happy slumber and to miserably move on their courses, maintaining the flow of time, fell at the then Vernal Equinox in Ayyaru, Ninurta's month of triumph.

And, as if to add insult to injury, Marduk was only third in line to be general for Enlil for this particular victory. As we learn from OECT 11 no. 69+70, Ninurta got his commission only after the legitimate son, the moongod Sîn, had proven inadequate to the task. Again, Ayyaru

summary 'Eternal return (Eliade),' *Wikipedia*: https://en.wikipedia.org/wiki/Eternal_return_(Eliade) (consulted August 22, 2018).

21 'The Elamite chariot which has no seat carries inside it the corpse of Enmešarra. The horses which are harnessed to it are the ghost of Anzû.' (Livingstone, *Court Poetry*, no. 39: 24-25).

22 OECT 11 no. 69+70 i 12'-16'.

23 'The king who stands in the chariot (drawn by Anzû's ghost and carrying the corpse of Enmešarra) is the warrior king, lord Ninurta. ... He hung a siege ladder on the wall of the house of Enmešarra; he woke the sleeper' (Livingstone, *Court Poetry*, no. 39: 25-26, 28-29).

24 Ephraim A. Speiser, 'The Myth of Zu,' in *Ancient Near Eastern Texts Relating to the Old Testament* (Ed. James B. Pritchard; 3rd edition; Princeton: Princeton University Press, 1969), 112-113, Assyrian Version Tablet 1 col. ii has Big Bird in a position of trust where he can observe for himself just what power the possessor of the Tablet of Destinies has at his command.

25 This will be entitled: 'The Sacred Divorce or Was Marduk Guilty?'

26 See above.

was the problem, since that month was, and still is in astrology, the month of exaltation of the moon.[27] As celebrated in the cult of Nippur, the moon god was the first to be elevated to kingship (meaning playing the role of general and military commander over the other gods) with Enmešarra's son Ennugi as vizier. Ninurta was not amused, arriving from his mountain and entering the Ešumeša in a rage on the 15th of Ayyaru.[28] His objections were overruled, and he had to settle for a heraldship on the 22nd.[29] As in the later *Enūma eliš*, the coming battle in the sky is supposed to be the victims' own fault. On the 23rd, vizier Ennugi is alleged to have been the one to engage in treachery, and the sons of Enmešarra thus take the blame for the opening of intergenerational hostilities. The need for a military response to this alleged threat brought warrior-god Ninurta to the forefront of action at the expense of dreamy Sîn.[30] The *akītu* of Ninurta was on the 24th when the defeated starry ancestors were dragged before a rejoicing Enlil.[31] As we know from a Neo-Assyrian cultic commentary, Ninurta's reward for his well-timed intervention was to receive kingship from the gods, his fathers, before going home to his mountain.[32] The real prize of victory, however, the precious Tablet of Destinies, went to Enlil on the 25th and 26th.[33]

But this was not the end of the story; on the 28th, Ninurta removed the nose-ropes of the Captive Gods.[34] Marduk, as we know, does the same with his eleven captives,[35] the future Zodiacal constellations minus Capricorn, the astral manifestation of Marduk's father Ea. The binding and loosing or, as later formulated, the death and rebirth of the stars, as well as their being forced to move across the sky was essential to the system. If Hydra was captured and bound and never released; if Leviathan was hunted once and that was it, there would be no more time.

On a day to day basis, Ninurta delegated the donkey work of herdsmanship to his pet planet Neberu who kept the stars, literally on track for the rest of the year.[36] Giving Marduk rather than Ninurta this time-creating victory had, therefore, also to involve appropriating this chief herding star, which is identified in the Babylonian Creation Epic as Marduk's star (*Enūma Eliš* VII 126-134). As an essential part of his creation of time, Marduk is supposed to have given Neberu the task of marking out the courses of the stars across the sky so that none of them could go wrong or stray (*Enūma Eliš* V 1-11). Neberu is, however, as Johannes Koch has shown, actually Ninurta's planet Mercury when stationary due east in the middle of the path of Anu.[37]

As Marduk's luck would have it, the progression of the equinoxes moved the all-important time-creating battle into Nisannu.[38] What this meant was that the defeat of Anu and the doing-

27 The exaltation of planets or the moon is referred to as *elûnum*.

28 OECT 11 no 69+ 70 i 17'-19'.

29 OECT 11 no 69+ 70 i 25'-27'

30 OECT 11 no 69+ 70 i 28'-31'.

31 OECT 11 no 69+ 70 i 32'-34'.

32 'The king, who wears on his head a golden crown from inside the temple and sits on a throne and whom they carry while going to the palace is Ninurta who avenged his father. The gods, his fathers, decorated him inside the Ekur, gave him sceptre, throne and rule, adorned him with the *melammu* of kingship and then he went out to the mountain' (Livingstone, *Court Poetry*, no. 39 r. 20-24).

33 OECT 11 no 69+ 70 i 35'-37'.

34 OECT 11 no 69+ 70 i 38'-39'.

35 *Enūma Eliš* IV 115-118.

36 JoAnn Scurlock, 'The World Encircling River,' in *Irrigation in Early States: New Directions* (ed. Stephanie Rost; Oriental Institute Seminars 13; Chicago: The Oriental Institute, forthcoming).

37 Johannes Koch, 'Der Mardukstern Neberu,' *Welt des Orients* 22 (1991): 48-72.

38 The change in the Ur calendar between Šu-Sîn 2 and Šu-Sîn 4 in which Nisannu rather than Ayyaru became the first

in of the older generation of gods could be celebrated simultaneously, making Marduk a great hero and pushing Ninurta into a dark corner, but still leaving Marduk as general for his father-in-law, Enlil. One would think that this would not raise any hackles, but raise them it did.

Instead of embracing the change, Nippur apparently refused to alter its cultic calendar to accommodate the new honors claimed by Marduk. As we have seen, OECT 11 no. 69+70 reveals that Nippur, long after the rise of Marduk to the head of the pantheon, was still performing rituals that honored Marduk as Enlil's general for the battle with Anu and Ninurta and not Marduk as King of the Gods and Lord of Time. Moreover, the most important of the ritually enacted events described in this late commentary may be lined up day by day with Tenney's Middle Babylonian offering list. For Ayyaru, we have the procession of the 10th to celebrate Enlil's marriage to Enmešarra's daughter, here identified with the Queen of Nippur.[39] The battlefield of the 15th is Ninurta's furious entry from the mountain to oppose Sîn's elevation to kingship.[40] The Great Festival of the 23rd is a celebration of Ninurta's intervention and great victory.[41] The *akītu* of the 24th is unarguably Ninurta's *akītu* when he presented the defeated gods to his father Enlil.[42] Finally, the temple, pure scepter and food portion of the king of the 25th/26th refer to Ninurta's accession to kingship but also Enlil's receipt of the Tablet of Destinies.[43]

It is hardly the case that Mesopotamians were allergic to alterations in established rituals, but there are two opposite ways of dealing with the new discoveries that upset previous understandings of the cosmos. You can acknowledge the new and go with the flow, which is what apparently happened in Assyria where they were aware that the Vernal Equinox first month of the year had once been in Ayyaru and was now in Nisannu.[44] Their response was to move the marriage of Annunitu to Enlil which was once celebrated in Ayyaru into Nisannu, re-imagining it as a Sacred Marriage between Ištar of Aššur and the Assyrian Enlil which made her Queen of Heaven.[45] Alternatively, you can dig in your heels and go on as if nothing had happened, which seems to have been Nippur's response.

It would be wrong to think that Nippur had its head in the sand. The problem is the same as the one the Pope had in the time of Galileo. Irrational Religion was not in inherent conflict with Rational Science as is often claimed. Instead, there was too much theology vested in the old science to allow for a change to the new science without a struggle. In the case of Galileo, the church had based important theological points about the imperfection of this earthly vale of tears by comparison with heavenly perfection on the mistaken assumptions of Aristotilean

month of the year (Cohen, *Festivals and Calendars*, 84-85) has been variously explained (Ibid. 83-85). The obvious explanation for this and the later adoption of the Nippur calendar (with its Nisannu first month) in the reign of Išbi-Erra in Southern Mesopotamia is the fact that these calendars began the year at the Vernal Equinox and that Ur was, unsurprisingly, the first to notice the progression of the equinoxes and made the appropriate adjustments. By the Middle Chronology. this would have been in ca. 2035 BCE at Ur with the rest of the south following suit sometime after 2017 BCE.

39 Tenney, 'Elevation of Marduk,' 164 obv. 13'.

40 Ibid., 164 obv. 14'.

41 Ibid., 164 obv. 15'.

42 Ibid., 164 obv. 16'.

43 Ibid., 164 obv. 17'-18'.

44 For references, see *The Assyrian Dictionary of the Oriental Institute of the University of Chicago* Z (Chicago: The Oriental Institute, 1961), 12 under *zagmukku*.

45 See JoAnn Scurlock, 'Feasting in the Garden of God', in this volume, above.

science that the earth was an irregular stationary body encircled by perfect spheres that moved gracefully by natural motion.[46]

Nippur had a similarly major theological investment in Ninurta as the original creator and maintainer of time. Ninurta was the city god of Nippur; his glory was their glory. Their hero and claim to cosmic significance could not simply be displaced in their theology by the upstart Amorite god of Babylon. So Nippur refused to grant Marduk's, from Babylon's point of view, reasonable request for primacy.

Fortunately, the Assyrian king Tukulti-Ninurta I's carrying off of Marduk to Aššur and Marduk's eventual return by a puppet Assyrian king installed, I would argue, by a Babylonian monarch quite desperate to get that statue back[47] provided the perfect opportunity to remedy the situation. It is long past time to bury forever the absurd notion that the AN.ŠAR of the *Enūma eliš* is anything other than the primordial form of the Assyrian national god, and to accept the obvious conclusion that the Babylonian Creation Epic was, in fact, a joint creation of Babylonia and Assyria, designed to allow for the return of Marduk to Babylon by Ninurta-tukulti-Aššur.[48] It is striking that the oldest textual witnesses to the so-called Babylonian epic are from the Assyrian cities of Aššur and Kalḫu.[49] In the middle Assyrian period, a smashed statue of Marduk was not languishing in a storehouse; instead, his cult was being practiced in rituals which survive in Assyria into the Neo-Assyrian period with the Assyrian king personally taking the part of Marduk to replace the now absent statue at key points.[50] Marduk could not simply go home without some mutually agreeable arrangements having been made.

In the story as we now have it, then, it was with the assent of the Assyrian national god Aššur, that Marduk was able to usurp Sumerian Enlil and Ninurta's places as Lord of Time and rightful Ruler of this World. Aššur was the transcendent Creator of the Universe, the Babylonian Creation Epic explains, but only at the very beginning, leaving the details of creation as well as the day to day running of the created world to other gods, eventually organized into a divine assembly under his leadership. Technically speaking, Marduk's annually celebrated victory over the sleepy older generation was still not his but that of Aššur, as is explicitly recognized

46 Brian Easlea, *Witch-hunting, Magic & the New Philosophy: An Introduction of the Debates of the Scientific Revolution 1450-1750* (Sussex: Harvester Press, 1980), 45-88.

47 For the intrigues with Babylonia, see A. Kirk Grayson, 'Ninurta-tukulti-Aššur, in *Reallexikon der Assyriologie* 9/7-8 (Berlin: Walter de Gruyter, 2001), 527. The fact that no years are given for Ninurta-tukulti-Aššur in the Assyrian king list does indeed suggest that whatever years were 'on his tablet' overlapped with those of the previous king or kings. We also know that he was at some point thrown out by his brother and ended up in Babylonia. Given the clear Babylonian interference represented by insulting letters sent to Assyria from Babylonia that praise him to the skies, I imagine him as a Babylonian puppet king installed, or more likely, re-installed with a view to recovering Marduk. This was not the sort of thing to celebrate, and both he and his brother were not, and did not deserve to be, considered true kings of Assyria.

48 A. Kirk Grayson, *Assyrian and Babylonian Chronicles* (Texts from Cuneiform Sources 5; Locust Valley: J.J. Augustin, 1975) no. 22 iv 12-13.

49 See Andrea Seri, 'Some Notes on *enūma eliš*,' *Journal of the American Oriental Society* 137 (2017): 834.

50 For the key part of the celebration with the statue in Middle Assyrian Aššur, see JoAnn Scurlock, 'Animal Sacrifice in Ancient Mesopotamian Religion,' in *A History of the Animal World in the Ancient Near East* (ed. Billie Jean Collins; Handbuch der Orientalistik 1/64; Leiden: Brill, 2002), 400-401. What is going on is the part in which a renewable relationship is established between the god (here Marduk) and a particular human community (the city of Aššur). After this has happened, we do not just wrap up the god and send him off somewhere for somebody else to keep. For the king as impersonating Marduk in the Neo-Assyrian ritual see especially: 'The king, who opens the vat in the race is Marduk, who [defeat]ed Tiamat with his penis' (Livingstone, *Court Poetry*, no. 37 obv. 18').

in the *Enūma eliš* where Marduk sets up not his own but Aššur's battle cry over the corpse of Tiamat.[51]

So Marduk was only King to Aššur's King of King's. But who cared? In Babylonia, Marduk was in charge. He could essentially do as he pleased, usurping functions claimed by other gods at will, even if had essentially to dethrone Enlil in the process. And the all-important task of having created and being charged with maintaining the flow of time belonged to him and to him alone.

Actually, Nippur cared, and a great deal. Fast forward to Sennacherib, grieving for his son, murdered by the Elamites with the connivance of Babylon. Are we to believe that proud Nippur, who complained so bitterly to Esarhaddon about having to contribute labor to the rebuilding of Babylon,[52] had nothing to do with its destruction? And for all those years that Marduk cooled his heels in the Netherworld waiting for someone to rebuild his city and temple, who but Ninurta, back to his old job that upstart Marduk stole from him, will have kept time flowing?

51 Enuma Eliš IV 125. For more details on the relationship of Aššur and Marduk, see JoAnn Scurlock, 'Chaoskampf Lost, Chaoskampf Regained: The Gunkel Hypothesis Revisited,' in *Creation and Chaos* (eds. JoAnn Scurlock and Richard Beal; Winona Lake: Eisenbrauns, 2013), 265-267.

52 JoAnn Scurlock, 'Marduk and his Enemies: City Rivalries in Southern Mesopotamia,' in *Organization, Representation, and Symbols of Power in the Ancient Near East: Proceedings of the 54th Rencontre Assyriologique Internationale at Würzburg 20-25 July 2008* (ed. Gernot Wilhelm, ed. Winona Lake: Eisenbraun, 2012), 369.

Index